ALGORITHMS
AND
DATA STRUCTURES

ALGORITHMS AND DATA STRUCTURES

Niklaus Wirth

ETH Zürich

Prentice-Hall, Inc., Englewood Cliffs, New Jersey 07632

Library of Congress Catalog Card Number: 85-61628
ISBN: 0-13-022005-1

Printed in the United States of America

10 9 8 7 6 5 4 3

ISBN 0-13-022005-1 01

Prentice-Hall International (UK) Limited, *London*
Prentice-Hall of Australia Pty. Limited, *Sydney*
Editora Prentice-Hall do Brasil, Ltda., *Rio de Janeiro*
Prentice-Hall Canada Inc., *Toronto*
Prentice-Hall Hispanoamericana, S.A., *Mexico*
Prentice-Hall of India Private Limited, *New Delhi*
Prentice-Hall of Japan, Inc., *Tokyo*
Prentice-Hall of Southeast Asia Pte. Ltd., *Singapore*
Whitehall Books Limited, *Wellington, New Zealand*

CONTENTS

PREFACE

In recent years the subject of *computer programming* has been recognized as a discipline whose mastery is fundamental and crucial to the success of many engineering projects and which is amenable to scientific treatement and presentation. It has advanced from a craft to an academic discipline. The initial outstanding contributions toward this development were made by E.W. Dijkstra and C.A.R. Hoare. Dijkstra's *Notes on Structured Programming* [1] opened a new view of programming as a scientific subject and intellectual challenge, and it coined the title for a "revolution" in programming. Hoare's *Axiomatic Basis of Computer Programming* [2] showed in a lucid manner that programs are amenable to an exacting analysis based on mathematical reasoning. Both these papers argue convincingly that many programmming errors can be prevented by making programmers aware of the methods and techniques which they hitherto applied intuitively and often unconsciously. These papers focused their attention on the aspects of composition and analysis of programs, or more explicitly, on the structure of algorithms represented by program texts. Yet, it is abundantly clear that a systematic and scientific approach to program construction primarily has a bearing in the case of large, complex programs which involve complicated sets of data. Hence, a methodology of programming is also bound to include all aspects of data structuring. *Programs,* after all, are concrete formulations of abstract *algorithms* based on particular representations and *structures of data*. An outstanding contribution to bring order into the bewildering variety of terminology and concepts on data structures was made by Hoare through his *Notes on Data Structuring* [3]. It made clear that decisions about structuring data cannot be made without knowledge of the algorithms applied to the data and that, vice versa, the structure and choice of algorithms often depend strongly on the structure of the underlying data. In short, the subjects of program composition and data structures are inseparably interwined.

Yet, this book starts with a chapter on data structure for two reasons. First, one has an intuitive feeling that data precede algorithms: you must have some objects before you can perform operations on them. Second, and this is the more immediate reason, this book assumes that the reader is familiar with the basic notions of computer programming. Traditionally and sensibly, however, introductory programming courses concentrate on algorithms operating on relatively simple structures of data. Hence, an introductorry chapter on data structures seems appropriate.

Throughout the book, and particularly in Chap. 1, we follow the theory and terminology expounded by Hoare and realized in the programming language PASCAL [4]. The essence of this theory is that data in the first instance represent abstractions of real phenomena and are preferably formulated as abstract structures not necessarily realized in common programming languages. In the process of program construction the data representation is gradually refined -- in step with the refinement of the algorithm -- to comply more and more with the constraints imposed by an available programming system [5]. We therefore postulate a number of basic building principles of data structures, called the *fundamental structures*. It is most important that they are constructs that are known to be quite easily implementable on actual computers, for only in this case can they be considered the true elements of an actual data representation, as the molecules emerging from the final step of refinements of the data description. They are the *record*, the *array* (with fixed size), and the *set*. Not surprisingly, these basic building principles correspond to mathematical notions that are fundamental as well.

A cornerstone of this theory of data structures is the distinction between fundamental and "advanced" structures. The former are the molecules -- themselves built out of atoms -- that are the components of the latter. Variables of a fundamental structure change only their value, but never their structure and never the set of values they can assume. As a consequence, the size of the store they occupy remains constant. "Advanced" structures, however, are characterized by their change of value *and* structure during the execution of a program. More sophisticated techniques are therefore needed for their implementation. The sequence appears as a hybrid in this classification. It certainly varies its length; but that change in structure is of a trivial nature. Since the sequence plays a truly fundamental role in practically all computer systems, its treatment is included in Chap. 1.

The second chapter treats *sorting algorithms*. It displays a variety of different methods, all serving the same purpose. Mathematical analysis of some of these algorithms shows the advantages and disadvantages of the methods, and it makes the programmer aware of the importance of analysis in the choice of good solutions for a given problem. The partitioning into methods for sorting arrays and methods for sorting files (often called internal and external sorting) exhibits the crucial influence of data representation on the choice of applicable algorithms and on their complexity. The space allocated to sorting would not be so large were it not for the fact that sorting constitutes an ideal vehicle for illustrating so many principles of programming and situations occurring in most other applications. It often seems that one could compose an entire programming course by deleting examples from sorting only.

Another topic that is usually omitted in introductory programming courses but one that plays an important role in the conception of many algorithmic solutions is recursion. Therefore, the third chapter is devoted to *recursive algorithms*. Recursion is shown to be a generalization of repetition (iteration), and as such it is an important and powerful concept in programming. In many programming tutorials, it is unfortunately exemplified by cases in which simple iteration would suffice. Instead, Chap. 3 concentrates on several examples of problems in which recursion allows for a most natural formulation of a solution, whereas use of iteration would lead to obscure and cumbersome programs. The class of *backtracking* algorithms emerges as an ideal application of recursion, but the most obvious candidates for the use of recursion are algorithms operating on data whose structure is defined recursively. These cases are treated in the last two chapters, for which the third chapter provides a welcome background.

Chapter 4 deals with *dynamic data structures,* i.e., with data that change their structure during the execution of the program. It is shown that the recursive data structures are an important subclass of the dynamic structures commonly used. Although a recursive definition is both natural and possible in these cases, it is usually not used in practice. Instead, the mechanism used in its implementation is made evident to the programmer by forcing him to use explicit reference or *pointer variables.* This book follows this technique and reflects the present state of the art: Chapter 4 is devoted to programming with pointers, to lists, trees and to examples involving even more complicated meshes of data. It presents what is often (and somewhat inappropriately) called *list processing.* A fair amount of space is devoted to tree organizations, and in particular to search trees. The chapter ends with a presentation of scatter tables, also called "hash" codes, which are oftern preferred to search trees. This provides the possibility of comparing two fundamentally different techniques for a frequently encountered application.

Programming is a *constructive* activity. How can a constructive, inventive activity be taught? One method is to crystallize elementary composition priciples out many cases and exhibit them in a systematic manner. But programming is a field of vast variety often involving complex intellectual activities. The belief that it could ever be condensed into a sort of pure recipe teaching is mistaken. What remains in our arsenal of teaching methods is the careful selection and presentation of master examples. Naturally, we should not believe that every person is capable of gaining equally much from the study of examples. It is the characteristic of this approach that much is left to the student, to his diligence and intuition. This is particularly true of the relatively involved and long example programs. Their inclusion in this book is not accidental. Longer programs are the prevalent case in practice, and they are much more suitable for exhibiting that elusive but essential ingredient called style and orderly structure. They are also meant to serve as exercises in the art of program reading, which too often is neglected in favor of program writing. This is a primary motivation behind the inclusion of larger programs as examples in their entirety. The reader is led through a gradual development of the program; he is given various snapshots in the evolution of a program, whereby this development becomes manifest as a *stepwise refinement* of the details. I consider it essential that programs are shown in final form with sufficient attention to details, for in programming, the devil hides in the details. Although the mere presentation of an algorithm's principle and its mathematical analysis may be stimulating and challenging to the academic mind, it seems dishonest to the engineering practitioner. I have therefore strictly adhered to the rule of presenting the final programs in a language in which they can actually be run on a computer.

Of course, this raises the problem of finding a form which at the same time is both machine executable and sufficiently machine independent to be included in such a text. In this respect, neither widely used languages nor abstract notations proved to be adequate.

The language Pascal provides an appropriate compromise; it had been developed with exactly this aim in mind, and it is therefore used throughout this book. The programs can easily be understood by programmers who are familiar with some other high-level language, such as ALGOL 60 or PL/1, because it is easy to understand the Pascal notation while proceeding through the text. However, this not to say that some proparation would not be beneficial. The book *Systematic Programming* [6] provides an ideal background because it is also based on the Pascal notation. The present book was, however, not intended as a manual on the language Pascal; there exist more appropriate texts for this purpose [7].

This book is a condensation -- and at the same time an elaboration -- of several courses on programming taught at the Federal Institute of Technology (ETH) at Zürich. I owe many ideas and views expressed in this book to discussions with my collaborators at ETH. In particular, I wish to thank Mr. H. Sandmayr for his careful reading of the manuscript, and Miss Heidi Theiler and D. Wirth for her care and patience in typing the text. I should also like to mention the stimulating influence provided by meetings of the Working Groups 2.1 and 2.3 of IFIP, and particularly the many memorable arguments I had on these occasions with E. W. Dijkstra and C.A.R. Hoare. Last but not least, ETH generously provided the environment and the computing facilities without which the preparation of this text would have been impossible.

Zürich, Aug. 1975 N. Wirth

1. In *Structured Programming.* O-.J. Dahl, E.W. Dijkstra, C.A.R. Hoare. F. Genuys, Ed. (New York; Academic Press, 1972), pp. 1-82.

2. In *Comm. ACM, 12,* No. 10 (1969), 576-83.

3. In *Structured Programming,* pp. 83-174.

4. N. Wirth. The Programming Language Pascal. *Acta Informatica, 1,* No. 1 (1971), 35-63.

5. N. Wirth. Program Development by Stepwise Refinement. *Comm. ACM, 14,* No. 4 (1971), 221-27.

6. N. Wirth. *Systematic Programming.* (Englewood Cliffs, N.J. Prentice-Hall, Inc., 1973.)

7. K. Jensen and N. Wirth, *PASCAL-User Manual and Report.* (Berlin, Heidelberg, New York; Springer-Verlag, 1974).

PREFACE TO THE 1986 EDITION

This new Edition incorporates many revisions of details and several changes of more significant nature. They were all motivated by experiences made in the ten years since the first Edition appeared. Most of the contents and the style of the text, however, have been retained. We briefly summarize the major alterations.

The major change which pervades the entire text concerns the programming language used to express the algorithms. Pascal has been replaced by *Modula-2*. Although this change is of no fundamental influence to the presentation of the algorithms, the choice is justified by the simpler and more elegant syntactic structures of Modula-2, which often lead to a more lucid representation of an algorithm's structure. Apart from this, it appeared advisable to use a notation that is rapidly gaining acceptance by a wide community, because it is well-suited for the development of large programming systems. Nevertheless, the fact that Pascal is Modula's ancestor is very evident and eases the task of a transition. The syntax of Modula is summarized in the Appendix for easy reference.

As a direct consequence of this change of programming language, Sect. 1.11 on the sequential file structure has been rewritten. Modula-2 does not offer a built-in file type. The revised Sect. 1.11 presents the concept of a *sequence* as a data structure in a more general manner, and it introduces a set of program modules that incorporate the sequence concept in Modula-2 specifically.

The last part of Chap. 1 is new. It is dedicated to the subject of *searching* and, starting out with linear and binary search, leads to some recently invented fast string searching algorithms. In this section in particular we use assertions and loop invariants to demonstrate the correctness of the presented algorithms.

A new section on *priority search trees* rounds off the chapter on dynamic data structures. Also this species of trees was unknown when the first Edition appeared. They allow an economical representation and a fast search of point sets in a plane.

The entire fifth chapter of the first Edition has been omitted. It was felt that the subject of compiler construction was somewhat isolated from the preceding chapters and would rather merit a more extensive treatment in its own volume.

Finally, the appearance of the new Edition reflects a development that has profoundly influenced publications in the last ten years: the use of computers and sophisticated algorithms to prepare and automatically typeset documents. This book was edited and laid out by the author with the aid of a *Lilith* computer and its document editor *Lara*. Without these tools, not only would the book become more costly, but it would certainly not be finished yet.

Palo Alto

N. Wirth

NOTATION

The following notations, adopted from publications of E.W. Dijkstra, are used in this book.

In logical expressions, the character & denotes conjunction and is pronounced as *and*. The character ~ denotes negation and is pronounced as *not*. Boldface **A** and **E** are used to denote the universal and existential quantifiers. In the following formulas, the left part is the notation used and defined here in terms of the right part. Note that the left parts avoid the use of the symbol "...", which appeals to the readers intuition.

$$\mathbf{A}i\text{: } m \leq i < n : P_i \quad \equiv \quad P_m \;\&\; P_{m+1} \;\&\; \dots \;\&\; P_{n-1}$$

The P_i are predicates, and the formula asserts that *for all* indices *i* ranging from a given value *m* to, but excluding a value *n*, P_i holds.

$$\mathbf{E}i\text{: } m \leq i < n : P_i \quad \equiv \quad P_m \text{ or } P_{m+1} \text{ or } \dots \text{ or } P_{n-1}$$

The P_i are predicates, and the formula asserts that *for some* indices *i* ranging from a given value *m* to, but excluding a value *n*, P_i holds.

$$\mathbf{S}i\text{: } m \leq i < n : x_i \quad = \quad x_m + x_{m+1} + \dots + x_{n-1}$$

$$\mathbf{MIN}\, i\text{: } m \leq i < n : x_i \quad = \quad \text{minimum}(x_m, x_{m+1}, \dots, x_{n-1})$$

$$\mathbf{MAX}\, i\text{: } m \leq i < n : x_i \quad = \quad \text{maximum}(x_m, x_{m+1}, \dots, x_{n-1})$$

Using quantifiers, the min and max operators can also be expressed as follows:

$$\mathbf{MIN}\, i\text{: } m \leq i < n : x_i \;=\; \min \;\equiv$$
$$(\mathbf{A}_i\text{: } m \leq i < n : \min \leq x_i) \;\&\; (\mathbf{E}_i\text{: } m \leq i < n : \min = x_i)$$

$$\mathbf{MAX}\, i\text{: } m \leq i < n : x_i \;=\; \max \;\equiv$$
$$(\mathbf{A}_i\text{: } m \leq i < n : \max \geq x_i) \;\&\; (\mathbf{E}_i\text{: } m \leq i < n : \max = x_i)$$

1 FUNDAMENTAL DATA STRUCTURES

1.1. INTRODUCTION

The modern digital computer was invented and intended as a device that should facilitate and speed up complicated and time-consuming computations. In the majority of applications its capability to store and access large amounts of information plays the dominant part and is considered to be its primary characteristic, and its ability to compute, i.e., to calculate, to perform arithmetic, has in many cases become almost irrelevant.

In all these cases, the large amount of information that is to be processed in some sense represents an *abstraction* of a part of reality. The information that is available to the computer consists of a selected set of *data* about the actual problem, namely that set that is considered relevant to the problem at hand, that set from which it is believed that the desired results can be derived. The data represent an abstraction of reality in the sense that certain properties and characteristics of the real objects are ignored because they are peripheral and irrelevant to the particular problem. An abstraction is thereby also a *simplification* of facts.

We may regard a personnel file of an employer as an example. Every employee is represented (abstracted) on this file by a set of data relevant either to the employer or to his accounting procedures. This set may include some identification of the employee, for example, his or her name and salary. But it will most probably not include irrelevant data such as the hair color, weight, and height.

In solving a problem with or without a computer it is necessary to choose an abstraction of reality, i.e., to define a set of data that is to represent the real situation. This choice must be guided by the problem to be solved. Then follows a choice of representation of this information. This choice is guided by the tool that is to solve the problem, i.e., by the facilities offered by the computer. In most cases these two steps are not entirely separable.

The choice of representation of data is often a fairly difficult one, and it is not uniquely determined by the facilities available. It must always be taken in the light of the operations that are to be performed on the data. A good example is the representation of numbers, which are themselves abstractions of properties of objects to be characterized. If addition is the only (or at least the dominant) operation to be performed, then a good way to represent the number n is to write n strokes. The addition rule on this representation is indeed very obvious and simple. The Roman numerals are based on the same principle of simplicity, and the adding rules are similarly straightforward for small numbers. On the other hand, the representation by Arabic numerals requires rules that are far from obvious (for small numbers) and they must be memorized. However, the situation is reversed when we consider either addition of large numbers or multiplication and division. The decomposition of these operations into simpler ones is much easier in the case of representation by Arabic numerals because of their systematic structuring principle that is based on positional weight of the digits.

It is generally known that computers use an internal representation based on binary digits (bits). This representation is unsuitable for human beings because of the usually large number of digits involved, but it is most suitable for electronic circuits because the two values 0 and 1 can be represented conveniently and reliably by the presence or absence of electric currents, electric charge, or magnetic fields.

From this example we can also see that the question of representation often transcends several levels of detail. Given the problem of representing, say, the position of an object, the first decision may lead to the choice of a pair of real numbers in, say, either Cartesian or polar coordinates. The second decision may lead to a floating-point representation, where every real number x consists of a pair of integers denoting a fraction f and an exponent e to a certain base (such that $x = f*2^e$). The third decision, based on the knowledge that the data are to be stored in a computer, may lead to a binary, positional representation of integers, and the final decision could be to represent binary digits by the direction of the magnetic flux in a magnetic storage device. Evidently, the first decision in this chain is mainly influenced by the problem situation, and the later ones are progressively dependent on the tool and its technology. Thus, it can hardly be required that a programmer decide on the number representation to be employed, or even on the storage device characteristics. These lower-level decisions can be left to the designers of computer equipment, who have the most information available on current technology with which to make a sensible choice that will be acceptable for all (or almost all) applications where numbers play a role.

In this context, the significance of *programming languages* becomes apparent. A programming language represents an abstract computer capable of interpreting the terms used in this language, which may embody a certain level of abstraction from the objects used by the actual machine. Thus, the programmer who uses such a higher-level language will be freed (and barred) from questions of number representation, if the number is an elementary object in the realm of this language.

The importance of using a language that offers a convenient set of basic abstractions common to most problems of data processing lies mainly in the area of reliability of the resulting programs. It is easier to design a program based on reasoning with familiar notions of numbers, sets, sequences, and repetitions than on bits, storage units, and jumps. Of course, an actual computer represents all data, whether numbers, sets, or sequences, as a large mass of bits. But this is irrelevant to the programmer as long as he or she does not have to worry about the details of representation of the chosen abstractions, and as long as he or she can rest assured that the corresponding representation chosen by the computer (or compiler) is reasonable for the stated purposes.

The closer the abstractions are to a given computer, the easier it is to make a representation choice for the engineer or implementor of the language, and the higher is the probability that a single choice will be suitable for all (or almost all) conceivable applications. This fact sets definite limits on the degree of abstraction from a given real computer. For example, it would not make sense to include geometric objects as basic data items in a general-purpose language, since their proper repesentation will, because of its inherent complexity, be largely dependent on the operations to be applied to these objects. The nature and frequency of these operations will, however, not be known to the designer of a general-purpose language and its compiler, and any choice the designer makes may be inappropriate for some potential applications.

In this book these deliberations determine the choice of notation for the description of algorithms and their data. Clearly, we wish to use familiar notions of mathematics, such as numbers, sets, sequences, and so on, rather than computer-dependent entities such as bitstrings. But equally clearly we wish to use a notation for which efficient compilers are known to exist. It is equally unwise to use a closely machine-oriented and machine-dependent language, as it is unhelpful to describe computer programs in an abstract notation that leaves problems of representation widely open. The programming language Pascal had been designed in an attempt to find a compromise between these extremes, and the language Modula-2 is the result of ten years' experience with Pascal [1-3]. It retains Pascal's basic concepts and incorporates some improvements and some extensions; it is used throughout this book [1-5]. Modula-2 has been successfully implemented on several computers, and it has been shown that the notation is sufficiently close to real machines that the chosen features and their representations can be clearly explained. The language is also sufficiently close to other languages, and hence the lessons taught here may equally well be applied in their use.

1.2. THE CONCEPT OF DATA TYPE

In mathematics it is customary to classify variables according to certain important characteristics. Clear distinctions are made between real, complex, and logical variables or between variables representing individual values, or sets of values, or sets of sets, or between functions, functionals, sets of functions, and so on. This notion of classification is equally if not more important in data processing. We will adhere to the principle that every constant, variable, expression, or function is of a certain *type*. This type essentially characterizes the set of values to which a constant belongs, or which can be assumed by a variable or expression, or which can be generated by a function.

In mathematical texts the type of a variable is usually deducible from the typeface without consideration of context; this is not feasible in computer programs. Usually there is one typeface available on computer equipment (i.e., Latin letters). The rule is therefore widely accepted that the associated type is made explicit in a *declaration* of the constant, variable, or function, and that this declaration textually precedes the application of that constant, variable, or function. This rule is particularly sensible if one considers the fact that a compiler has to make a choice of representation of the object within the store of a computer. Evidently, the amount of storage allocated to a variable will have to be chosen according to the size of the range of values that the variable may assume. If this information is known to a compiler, so-called dynamic storage allocation can be avoided. This is very often the key to an efficient realization of an algorithm. The primary characteristics of the concept of type that is used throughout this text, and that is embodied in the programming language Modula-2, are the following [1-2]:

1. A data type determines the set of values to which a constant belongs, or which may be assumed by a variable or an expression, or which may be generated by an operator or a function.

2. The type of a value denoted by a constant, variable, or expression may be derived from its form or its declaration without the necessity of executing the computational process.

3. Each operator or function expects arguments of a fixed type and yields a result of a fixed type. If an operator admits arguments of several types (e.g., + is used for addition of both integers and real numbers), then the type of the result can be determined from specific language rules.

As a consequence, a compiler may use this information on types to check the legality of various constructs. For example, the mistaken assignment of a Boolean (logical) value to an arithmetic variable may be detected without executing the program. This kind of redundancy in the program text is extremely useful as an aid in the development of programs, and it must be considered as the primary advantage of good high-level languages over machine code (or symbolic assembly code). Evidently, the data will ultimately be represented by a large number of binary digits, irrespective of whether or not the program had initially been conceived in a high-level language using the concept of type or in a typeless assembly code. To the computer, the store is a homogeneous mass of bits without apparent structure. But it is exactly this abstract structure which alone is enabling human programmers to recognize meaning in the monotonous landscape of a computer store.

The theory presented in this book and the programming language Modula-2 specify certain methods of defining data types. In most cases new data types are defined in terms of previously defined data types. Values of such a type are usually conglomerates of component values of the previously defined constituent types, and they are said to be *structured.* If there is only one constituent type, that is, if all components are of the same constituent type, then it is known as the *base type.* The number of distinct values belonging to a type T is called its *cardinality.* The cardinality provides a measure for the amount of storage needed to represent a variable x of the type T, denoted by *x: T.*

Since constituent types may again be structured, entire hierarchies of structures may be built up, but, obviously, the ultimate components of a structure are atomic. Therefore, it is necessary that a notation is provided to introduce such primitive, unstructured types as well. A straightforward method is that of *enumerating* the values that are to constitute the type. For example in a program concerned with plane geometric figures, we may introduce a primitive type called *shape,* whose values may be denoted by the identifiers *rectangle, square, ellipse, circle.* But apart from such programmer-defined types, there will have to be some standard, predefined types. They usually include numbers and logical values. If an ordering exists among the individual values, then the type is said to be ordered or *scalar.* In Modula-2, all unstructured types are ordered; in the case of explicit enumeration, the values are assumed to be ordered by their enumeration sequence.

With this tool in hand, it is possible to define primitive types and to build conglomerates, structured types up to an arbitrary degree of nesting. In practice, it is not sufficient to have only one general method of combining constituent types into a structure. With due regard to practical problems of representation and use, a general-purpose programming language must offer several *methods of structuring.* In a mathematical sense, they are equivalent; they differ in the operators available to select components of these structures. The basic structuring methods presented here are the *array,* the *record,* the *set,* and the *sequence.* More complicated structures are not usually defined as static types, but are instead dynamically generated during the execution of the program, when they may vary in size *and* shape. Such structures are the subject of Chap. 4 and include lists, rings, trees, and general, finite graphs.

Variables and data types are introduced in a program in order to be used for computation. To this end, a set of operators must be available. For each standard data type a programming languages offers a certain set of primitive, standard operators, and likewise with each structuring method a distinct operation and notation for selecting a component. The task of composition of operations is often considered the heart of the art of programming. However, it will become evident that the appropriate composition of data is equally fundamental and essential.

The most important basic operators are *comparison* and *assignment,* i.e., the test for equality (and for order in the case of ordered types), and the command to enforce equality. The fundamental difference between these two operations is emphasized by the clear distinction in their denotation throughout this text.

Test for equality: x = y (an expression with value TRUE or FALSE)
Assignment to x: x := y (a statement making x equal to y)

These fundamental operators are defined for most data types, but it should be noted that their execution may involve a substantial amount of computational effort, if the data are large and highly structured.

For the standard primitive data types, we postulate not only the availability of assignment and comparison, but also a set of operators to create (compute) new values. Thus we introduce the standard operations of arithmetic for numeric types and the elementary operators of propositional logic for logical values.

1.3. PRIMITIVE DATA TYPES

A new, primitive type is definable by enumerating the distinct values belonging to it. Such a type is called an *enumeration type.* Its definition has the form

$$\text{TYPE } T = (c_1, c_2, \dots , c_n) \tag{1.1.}$$

T is the new type identifier, and the c_i are the new constant identifiers. The cardinality of T is card(T) = n.

EXAMPLES

```
TYPE shape = (rectangle, square, ellipse, circle)
TYPE color = (red, yellow, green)
TYPE sex = (male, female)
TYPE BOOLEAN = (FALSE, TRUE)
TYPE weekday = (Monday, Tuesday, Wednesday, Thursday, Friday,
                Saturday, Sunday)
TYPE currency = (franc, mark, pound, dollar, shilling, lira, guilder,
                krone, ruble, cruzeiro, yen)
TYPE destination = (hell, purgatory, heaven)
TYPE vehicle = (train, bus, automobile, boat, airplane)
TYPE rank = (private, corporal, sergeant, lieutenant, captain, major,
            colonel, general)
TYPE object = (constant, type, variable, procedure, module)
TYPE structure = (array, record, set, sequence)
TYPE condition = (manual, unloaded, parity, skew)
```

The definition of such types introduces not only a new type identifier, but at the same time the set of identifiers denoting the values of the new type. These identifiers may then be used as constants throughout the program, and they enhance its understandability considerably. If, as an example, we introduce variables s, d, r, and b.

```
VAR s: sex
VAR d: weekday
VAR r: rank
VAR b: BOOLEAN
```

then the following assignment statements are possible:

```
s := male
d := Sunday
r := major
b := TRUE
```

Evidently, they are considerably more informative than their counterparts

```
s := 1   d := 7   r := 6   b := 2
```

which are based on the assumption that c, d, r, and b are defined as integers and that the constants are mapped onto the natural numbers in the order of their enumeration. Furthermore, a compiler can check against the inconsistent use of operators. For example, given the declaration of s above, the statement $s := s+1$ would be meaningless.

If, however, we recall that enumerations are ordered, then it is sensible to introduce operators that generate the successor and predecessor of their argument. We therefore postulate the following standard operators, which assign to their argument its successor and predecessor respectively:

INC(x)   DEC(x)   (1.2)

1.4. STANDARD PRIMITIVE TYPES

Standard primitive types are those types that are available on most computers as built-in features. They include the whole numbers, the logical truth values, and a set of printable characters. On many computers fractional numbers are also incorporated, together with the standard arithmetic operations. We denote these types by the identifiers

INTEGER, CARDINAL, REAL, BOOLEAN, CHAR

The type INTEGER comprises a subset of the whole numbers whose size may vary among individual computer systems. If a computer uses n bits to represent an integer in two's complement notation, then the admissible values x must satisfy $-2^{n-1} \leq x < 2^{n-1}$. It is assumed that all operations on data of this type are exact and correspond to the ordinary laws of arithmetic, and that the computation will be interrupted in the case of a result lying outside the representable subset. This event is called *overflow*. The standard operators are the four basic arithmetic operations of addition (+), subtraction (-), multiplication (*), and division (/, DIV).

We shall here distinguish between *Eulerian integer arithmetic* and modulo arithmetic. In the former, division is denoted by a slash, and the remainder of division by the operator REM. Let the quotient be q = m/n and the remainder r = m REM n. Then the relations

$$q*n + r = m \quad \text{and} \quad 0 \leq \text{ABS}(r) < \text{ABS}(n) \qquad (1.3)$$

always hold. The sign of the remainder is the same as that of the dividend (or the remainder is zero). Hence, Eulerian integer division is symmetric with respect to zero, as expressed by the following equalities:

$$(-m)/n = m/(-n) = -(m/n)$$

Examples:

```
 31 /  10 =  3        31 REM  10  =  1
-31 /  10 = -3       -31 REM  10  = -1
 31 / -10 = -3        31 REM -10  =  1
-31 / -10 =  3       -31 REM -10  = -1
```

In modulus (or congruence) arithmetic, the value m MOD n is actually a congruence class, i.e. a set of integers rather than a single number. This set contains all integers m-Qn for arbitrary integers Q. Obviously, classes can be represented by a specific member, e.g. by their smallest nonnegative member. Hence we define R = m MOD n, and at the same time Q = m DIV n such that the following relations are satisfied:

$$Q*n + R = m \quad \text{and} \quad 0 \leq R < n \qquad (1.4)$$

Examples:

```
 31 DIV 10 =  3        31 MOD 10  = 1
-31 DIV 10 = -4       -31 MOD 10  = 9
```

We note that dividing by 10^n can be achieved by merely shifting the decimal digits n places to the right and thereby ignoring the lost digits. The same method applies, if numbers are represented in binary instead of decimal form. If two's complement representation is used (as in practically all modern computers), then the shifts implement a division as defined by the above DIV operator (but not by the / operator). Moderately sophisticated compilers will therefore represent an operation of the form m DIV 2^n by a fast shift operation, whereas this shortcut is not permitted in expressions of the form $m / 2^n$.

If a variable, such as a counter, is known to assume no negative values, then this fact can be expressed by the use of a further standard type called CARDINAL. On a computer using n bits to represent (unsigned) integers, variables of this type can be assigned values x which satisfy $0 \leq x < 2^n$.

The type REAL denotes a subset of the real numbers. Whereas arithmetic with operands of the types INTEGER or CARDINAL is assumed to yield exact results, arithmetic on values of type REAL is permitted to be inaccurate within the limits of round-off errors caused by computation on a finite number of digits. This is the principal reason for the explicit distinction between the types INTEGER and REAL, as it is made in most programming languages.

The two values of the standard type BOOLEAN are denoted by the identifiers TRUE and FALSE. The Boolean operators are the logical conjunction, disjunction, and negation whose values are defined in Table 1.1. The logical conjunction is denoted by the symbol AND (or &), the logical disjunction by OR, and negation by NOT (or ~). Note that comparisons are operations yielding a result of type BOOLEAN. Thus, the result of a comparison may be assigned to a variable, or it may be used as an operand of a logical operator in a Boolean expression. For instance, given Boolean variables p and q and integer variables x = 5, y = 8, z = 10, the two assignments

```
p := x = y
q := (x < y) & (y < z)
```

yield p = FALSE and q = TRUE.

p	q	p AND q	p OR q	NOT p
TRUE	TRUE	TRUE	TRUE	FALSE
TRUE	FALSE	TRUE	FALSE	FALSE
FALSE	TRUE	TRUE	FALSE	TRUE
FALSE	FALSE	FALSE	FALSE	TRUE

Table 1.1 Boolean Operators.

The Boolean operators AND and OR have an additional property in Modula-2 (and in some other languages), which distinguishes them from other dyadic operators. Whereas, for example, the sum x+y is not defined, if either x or y is undefined, the conjunction p&q is defined even if q is undefined, provided that p is FALSE. This conditionality is an important and useful property. The exact definition of AND and OR is therefore given by the following equations:

$$\begin{aligned} p \text{ AND } q &= \text{if } p \text{ then } q \text{ else FALSE} \\ p \text{ OR } q &= \text{if } p \text{ then TRUE else } q \end{aligned} \qquad (1.5)$$

The standard type CHAR comprises a set of printable characters. Unfortunately, there is no generally accepted standard character set used on all computer systems. Therefore, the use of the predicate "standard" may in this case be almost misleading; it is to be understood in the sense of "standard on the computer system on which a certain program is to be executed."

The character set defined by the International Standards Organization (ISO), and particularly its American version ASCII (American Standard Code for Information Interchange) is the most widely accepted set. The ASCII set is therefore tabulated in Appendix A. It consists of 95 printable (graphic) characters and 33 control characters, the latter mainly being used in data transmission and for the control of printing equipment.

In order to be able to design algorithms involving characters (i.e., values of type CHAR) that are system independent, we should like to be able to assume certain properties of character sets as binding, namely,

1. The type CHAR contains the 26 capital Latin letters, the 26 lower-case letters, the 10 Arabic digits, and a number of other graphic characters, such as punctuation marks.
2. The subsets of letters and digits are ordered and contiguous, i.e.,

$$\begin{aligned} &(\text{"A"} \leq x) \,\&\, (x \leq \text{"Z"}) \rightarrow x \text{ is a capital letter} \\ &(\text{"a"} \leq x) \,\&\, (x \leq \text{"z"}) \rightarrow x \text{ is a lower-case letter} \\ &(\text{"0"} \leq x) \,\&\, (x \leq \text{"9"}) \rightarrow x \text{ is a digit} \end{aligned} \qquad (1.6)$$

3. The type CHAR contains a non-printing, blank character that may be used as a separator.

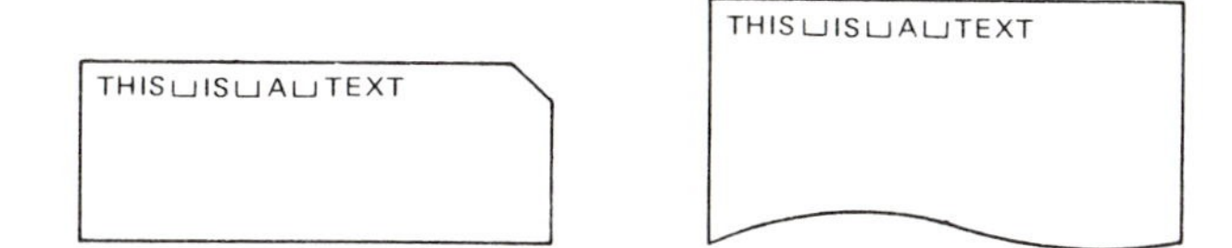

Fig. 1.1 Representations of a text.

The availability of two standard type transfer functions between the types CHAR and CARDINAL is particularly important in the quest to write programs in a machine independent form. We will call them ORD(ch), denoting the ordinal number of *ch* in the character set, and CHR(i), denoting the character with ordinal number i. Thus, CHR is the inverse function of ORD, and vice versa, that is,

$$\begin{aligned} ORD(CHR(i)) &= i \quad \text{(if CHR(i) is defined)} \\ CHR(ORD(c)) &= c \end{aligned} \tag{1.7}$$

Furthermore, we postualte a standard function CAP(ch). Its value is defined as the capital letter corresponding to ch, provided ch is a letter.

$$\begin{aligned} &\text{ch is a lower-case letter} \rightarrow \text{CAP(ch)} = \text{corresponding capital letter} \\ &\text{ch is a capital letter} \rightarrow \text{CAP(ch)} = \text{ch} \end{aligned} \tag{1.8}$$

1.5. SUBRANGE TYPES

It is often the case that a variable assumes values of a certain type within a specific interval only. This can be expressed by defining the variable to be of a subrange type according to the form

```
TYPE T = [min .. max]                                    (1.9)
```

where *min* and *max* are expressions specifying the limits of the interval. Note that these expressions must contain constants only as operands.

EXAMPLES

```
TYPE year     = [1900 .. 1999]
TYPE letter   = ["A" .. "Z"]
TYPE digit    = ["0" .. "9"]
TYPE officer  = [lieutenant .. general]
TYPE index    = [0 .. 2*N-1]
```

Given the variables

```
VAR y: year
VAR L: letter
```

the assignments y := 1984 and L := "L" are permissible, but y := 1291 and L := "9" are not. However, the legality of such assignments can be verified without execution of the program only if the assigned value is a constant. The admissibility of assignments of the kind *y := i* and *L := c*, where i is of type INTEGER and c is of type CHAR, cannot be checked through a mere textual scan by the compiler. In practice, systems performing these checks during program execution have proved enormously valuable in program development. Their utilization of redundant information to detect possible errors is again a prime motivation for using high-level languages.

1.6. THE ARRAY STRUCTURE

The array is probably the most widely known data structure; in some languages it is even the only one available. An array consists of components which are all of the same type, called its *base type*; it is therefore called a *homogeneous* structure. The array is also a so-called *random-access* structure; all components can be selected at random and are equally accessible. In order to denote an individual component, the name of the entire structure is augmented by the so-called *index* selecting the component. This index is to be a value of the type defined as the *index type* of the array. The definition of an array type T therefore specifies both a base type T0 and an index type TI.

```
TYPE T = ARRAY  TI  OF T0                                    (1.10)
```

EXAMPLES

```
TYPE Row    = ARRAY [1 .. 5] OF REAL
TYPE Card   = ARRAY [1 .. 80] OF CHAR
TYPE alfa   = ARRAY [0 .. 15] OF CHAR
```

A particular value of a variable

```
VAR x: Row
```

with all components satisfying the equation $x_i = 2^{-i}$, may be visualized as shown in Fig. 1.2.

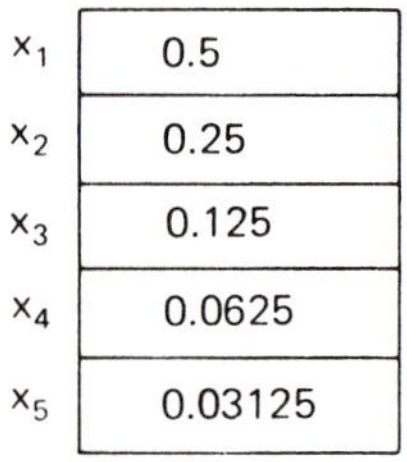

x_1	0.5
x_2	0.25
x_3	0.125
x_4	0.0625
x_5	0.03125

Fig. 1.2 Array of type Row.

An individual component of an array can be selected by an index. Given an array variable x, we denote an array selector by the array name followed by the respective component's index i, and we write x_i or x[i]. Because of the first, conventional notation, a component of an array component is therefore also called a *subscripted variable.*

The common way of operating with arrays, particularly with large arrays, is to selectively update single components rather than to construct entirely new structured values. This is expressed by considering *an array variable as an array of component variables* and by permitting assignments to selected components, such as for example x[i] := 0.125. Although selective updating causes only a single component value to change, from a conceptual point of view we must regard the entire composite value as having changed too.

The fact that array indices, i.e., names of array components, must be of a defined (scalar) data type has a most important consequence: *indices may be computed.* A general index expression may be substituted in place of an index constant; this expression is to be evaluated, and the result identifies the selected component. This generality not only provides a most significant and powerful programming facility, but at the same time it also gives rise to one of the most frequently encountered programming mistakes: The resulting value may be outside the interval specified as the range of indices of the array. We will assume that decent computing systems provide a warning in the case of such a mistaken access to a non-existent array component.

The cardinality of a structured type, i. e. the number of values belonging to this type, is the product of the cardinality of its components. Since all components of an array type T are of the same base type T0, we obtain, given index type TI

$$card(T) = card(T0)^{card(TI)} \tag{1.11}$$

Constituents of array types may themselves be structured. An array variable whose components are again arrays is called a *matrix.* For example,

```
M: ARRAY [1 .. 10] OF Row
```

is an array consisting of ten components (rows), each constisting of five components of type REAL, and is called a 10×5 matrix with real components. Selectors may be concatenated accordingly, such that M_{ij} and M[i][j] denote the j th component of row M_i, which is the i th component of M. This is usually abbreviated as M[i, j] and in the same spirit the declaration

```
M: ARRAY[1 .. 10] OF ARRAY[1 .. 5] OF REAL
```

can be written more concisely as

```
M: ARRAY[1 .. 10], [1 .. 5] OF REAL.
```

If a certain operation has to be performed on all components of an array or on adjacent components of a section of the array, then this fact may conveniently be emphasized by using the FOR satement, as shown in the following examples for computing the sum and for finding the maximal element of an array declared as

```
VAR a: ARRAY [0 .. N-1] OF INTEGER

sum := 0;
FOR i := 0 TO N-1 DO sum := a[i] + sum END

k := 0; max := a[0];
FOR i := 1 TO N-1 DO
  IF max < a[i] THEN k := i; max := a[k] END
END.
```

In a further example, assume that a fraction f is represented in its decimal form with k-1 digits, i.e., by an array d such that

$$f = Si: 1 \le i < k: \ d_i * 10^{-i}.$$

Now assume that we wish to divide f by 2. This is done by repeating the familiar division operation for all k-1 digits d_i, starting with i=1. It consists of dividing each digit by 2 taking into account a possible carry from the previous position, and of retaining a possible remainder r for the next position:

```
r := 10*r +d[i];  d[i] := r DIV 2;  r := r MOD 2
```

This procedure is applied in Program 1.1 to compute a table of negative powers of 2. The repetition of halving to compute $2^{-1}, 2^{-2}, \ldots, 2^{-N}$ is again appropriately expressed by a FOR statement, thus leading to a nesting of two FOR statements.

```
MODULE Power;
  (*compute decimal representation of negative powers of 2*)
  FROM InOut IMPORT Write, WriteLn;
  CONST N = 10;
  VAR i, k, r: CARDINAL;
    d: ARRAY [1 .. N] OF CARDINAL;
BEGIN
  FOR k := 1 TO N DO
    Write("."); r := 0;
    FOR i := 1 TO k-1 DO
      r := 10*r + d[i]; d[i] := r DIV 2; r := r MOD 2;
      Write(CHR(d[i] + ORD("0")))
    END ;
    d[k] := 5; Write("5"); WriteLn
  END
END Power.
```

Program 1.1 Compute powers of 2.

The resulting output for n = 10 is

```
.5
.25
.125
.0625
.03125
.015625
.0078125
.00390625
.001953125
.0009765625
```

1.7. THE RECORD STRUCTURE

The most general method to obtain structured types is to join elements of arbitrary types, that are possibly themselves structured types, into a compound. Examples from mathematics are complex numbers, composed of two real numbers, and coordinates of points, composed of two or more numbers according to the dimensionality of the space spanned by the coordinate system. An example from data processing is describing people by a few relevant characteristics, such as their first and last names, their date of birth, sex, and marital status.

In mathematics such a compound type is called the *Cartesian product* of its constituents types. This stems from the fact that the set of values defined by this compound type consists of all possible combinations of values, taken one from each set defined by each constituent type. Thus, the number of such combinations, also called *n-tuples,* is the product of the number of elements in each constituent set, that is, the cardinality of the compound type is the product of the cardinalities of the constituent types.

In data processing, composite types, such as descriptions of persons or objects, usually occur in files or data banks and record the relevant characteristics of a person or object. The word *record* has therefore become widely accepted to describe a compound of data of this nature, and we adopt this nomenclature in preference to the term Cartesian product. In general, a record type T with components of the types $T_1, T_2, \dots, T_n$ is defined as follows:

```
TYPE T =      RECORD s1 : T1;
                     s2 : T2;
                     ...                                    (1.12)
                     sn : Tn
              END
```

$$card(T) = card(T_1) * card(T_2) * \dots * card(T_n)$$

EXAMPLES

```
TYPE Complex = RECORD re: REAL;
                      im: REAL
               END

TYPE Date =    RECORD day: [1 .. 31];
                      month: [1 .. 12];
                      year: [1 .. 2000]
               END

TYPE Person =  RECORD name, firstname: alfa;
                      birthdate: Date;
                      sex: (male, female);
                      marstatus: (single, married, widowed, divorced)
               END
```

We may visualize particular, record-structured values of, for example, the variables

z: Complex
d: Date
p: Person

as shown in Fig. 1.3.

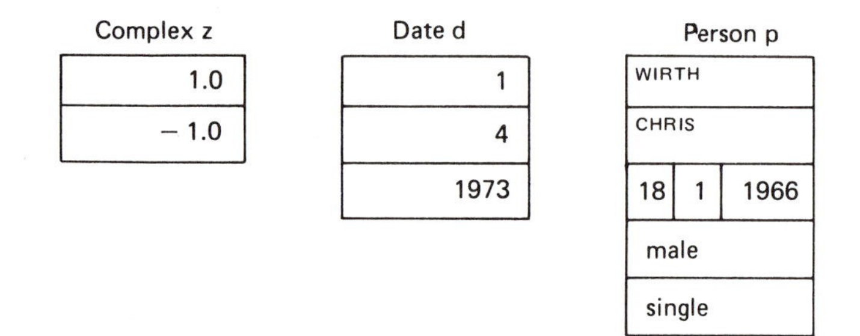

Fig. 1.3 Records of the types Complex, Date, and Person.

The identifiers $s_1, s_2, \dots, s_n$ introduced by a record type definition are the names given to the individual components of variables of that type. As components of records are called *fields*, the names are called *field identifiers*. They are used in record selectors applied to record structured variables. Given a variable x: T, its i-th field is denoted by $x.s_i$. Selective updating of x is achieved by using the same selector denotation on the left side in an assignment statement:

$$x.s_i := e$$

where e is a value (expression) of type T_i. Given, for example, the record variables z, d, and p declared above, the following are selectors of components:

```
z.im               (of type REAL)
d.month            (of type [1 .. 12])
p:name             (of type alfa)
p.birthdate        (of type Date)
p.birthdate.day    (of type [1 .. 31])
```

The example of the type *Person* shows that a constituent of a record type may itself be structured. Thus, selectors may be concatenated. Naturally, different structuring types may also be used in a nested fashion. For example, the i-th component of an array *a* being a component of a record variable *r* is denoted by r.a[i], and the component with the selector name *s* of the i-th record structured component of the array *a* is denoted by a[i].s.

It is a characteristic of the Cartesian product that it contains all combinations of elements of the constituent types. But it must be noted that in practical applications not all of them may be meaningful. For instance, the type *Date* as defined above includes the 31st April as well as the 29th February 1985, which are both dates that never occurred. Thus, the definition of this type does not mirror the actual situation entirely correctly; but it is close enough for practical purposes, and it is the responsibility of the programmer to ensure that meaningless values never occur during the execution of a program.

The following short excerpt from a program shows the use of record variables. Its purpose is to count the number of persons represented by the array variable *family* that are both female and single:

```
VAR count: CARDINAL;
   family: ARRAY [1 .. N] OF Person;
   count := 0;

FOR i := 1 TO N DO
   IF (family[i].sex = female) & (family[i].marstatus = single) THEN
     count := count + 1
   END
END
```

A notational variant of the above statement uses a construct that is called a *with-statement:*

```
FOR i := 1 TO N DO
  WITH family[i] DO
    IF (sex = female) & (marstatus = single) THEN
      count := count + 1
    END
  END
END
```

The meaning of "WITH r DO s END" is that field identifiers of the type of the variable r may be used without prefix within the statement s, and that they are taken to refer to the variable r. The with-statement thus serves to abbreviate the program text as well as to prevent frequent reevaluation of the storage address of the indexed component family[i].

The record structure and the array structure have the common property that both are *random-access* structures. The record is more general in the sense that there is no requirement that all constituent types must be identical. In turn, the array offers greater flexibility by allowing its component selectors to be computable values (expressions), whereas the selectors of record components are field identifiers declared in the record type definition.

1.8. VARIANTS OF RECORD STRUCTURES

In practice, it is often convenient and natural to consider two types simply as variants of the same type. For example, a type *Coordinate* of the preceding section may be regarded as the union of its two variants of Cartesian and polar coordinates whose constituents are (a) two lengths and (b) a length and an angle, respectively. In order to identify the variant actually assumed by a variable, we introduce a third component, the *type discriminator* or *tag field*.

```
TYPE CoordMode = (Cartesian, polar);
TYPE Coordinate =
        RECORD
          CASE kind: CoordMode OF
           Cartesian:  x, y: REAL |
           polar:      r: REAL; phi: REAL
          END
        END
```

Here, the name of the tag field is *kind*, and the names of the coordinates are either x and y in the case of a Cartesian value, or they are r and *phi* in the case of a polar value. The set of values denoted by this type Coordinate is the union of the two types

T_1 = (x, y: REAL)

T_2 = (r: REAL; phi REAL)

and its cardinality is the sum of the cardinalities of T_1 and T_2.

$$\mathrm{card}(T) = \mathrm{card}(T_1) + \mathrm{card}(T_2) \qquad (1.13)$$

Quite often, however, the variants are not entirely distinct types, but rather two types with partly identical components. It is this predominant situation which gave rise to the term *variant* record structure. An example is that of the type *Person*, defined in the preceding section in which the relevant characteristics to be recorded in a file depend on the sex of the person. For example, for a male, his weight and whether or not he is bearded may be regarded as relevant in a particular situation, but for a female, the three characteristic measurements may be considered as relevant data. The following is a type definition which reflects these circumstances:

```
TYPE Person =
  RECORD name, firstname: alfa;
    birthdate: Date;
    marstatus: (single, married, widowed, divorced);
    CASE s: sex OF
     male:   weight: REAL; bearded: BOOLEAN |
     female: size: ARRAY [1 .. 3] OF INTEGER
    END
```

END

The general form of a variant record type definition with m variants is

```
TYPE T =                                                              (1.14)
  RECORD s1: T1; s2: T2; ... ; sn-1: Tn-1;
    CASE sn: Tn OF
      v1: s11: T11; ... ; s1,n1: T1,n1 |
      v2: s21: T21; ... ; s2,n2: T2,n2 |
      . . .
      vm: sm,1: Tm,1; ... ; sm,nm: Tm,nm
    END
  END
```

The s_i (for i = 1 ... n-1) are the field identifiers belonging to the *common part* of the record, the s_{ij} are the selector names of the fields belonging to the *variant part,* and s_n is the name of the descriminating *tag field* with type T_n. The constants $v_1, v_2, \ldots, v_m$ denote the values of the (scalar) tag field type. Each variant i has n_i fields in its variant part. A variable x of type T consists of the components

$$x.s_1, x.s_2, \ldots, x.s_n, x.s_{k,1}, \ldots, x.s_{k,n_k}$$

if and only if the (current) value of $x.s_n = v_k$. Therefore, the use of a component selector $x.s_{k,h}$ $(1 \leq h \leq n_k)$ when $x.s_n \neq v_k$ must be regarded as a serious programming mistake and (in reference to the type Person defined above) amounts to asking wether or not a woman is bearded or (in the case of selective updating) ordering her to be so. When using variant records, utmost care is therefore required, and corresponding operations on the individual variants are best grouped into a selective statement, the so-called *case statement,* whose structure closely mirrors that of the variant record type definition.

```
CASE x.sn OF
    v1 : S1;
    v2 : S2;
    . . .                                                             (1.15)
    vm : Sm
END
```

S_k stands for the statement catering for the case that x assumes the form of variant k, i.e., it is selected for execution if and only if the tag field $x.s_n$ has the value v_k. As a consequence, it is fairly easy to safeguard against the misuse of selector names by verifying that each S_k contains only selectors

$$x.s_1 \ldots x.s_{n-1} \quad \text{and} \quad x.s_{k,1} \ldots x.s_{k,n_k}$$

The following short sample program illustrates the use of a case statement. Its purpose is to compute the distance *d* between two points A and B given the variables *a* and *b* of the variant record type *Coordinate*. The computational procedure differs according to the four possible combinations of Cartesian and polar coordinates (see Fig. 1.4).

```
CASE a.kind OF
  Cartesian: CASE b.kind OF
               Cartesian: d := sqrt(sqr(a.x-b.x) + sqr(a.y-b.y)) |
               Polar:     d := sqrt(sqr(a.x - b.r*cos(b.phi) + sqr(a.y - b.r*sin(b.phi))
             END |
  Polar:     CASE b.kind OF
               Cartesian: d := sqrt(sqr(a.r*cos(a.phi) - b.x) + sqr(a.r*sin(a.phi) - b.y)) |
               Polar:     d := sqrt(sqr(a.r + sqr(b.r) - 2*a.r*cos(a.phi-b.phi))
             END
END
```

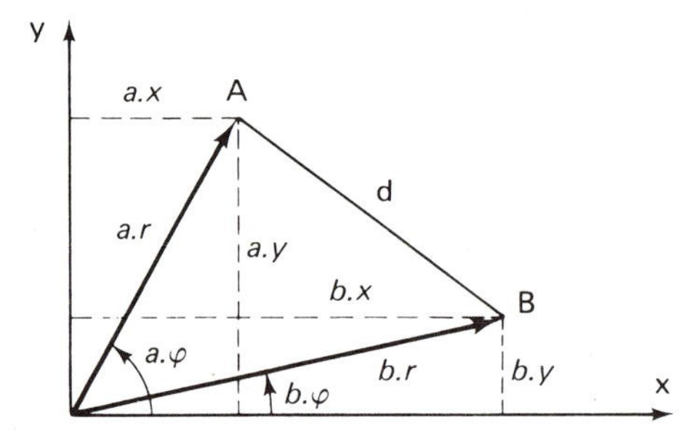

Fig. 1.4 Cartesian and polar coordinates.

1.9. THE SET STRUCTURE

The third fundamental data structure is the *set* structure. It is defined by the following declaration pattern:

TYPE T = SET OF T_0 (1.16)

The possible values of a variable x of type T are sets of elements of T_0. The set of all subsets of elements of a set T_0 is called the *powerset* of T_0. The type T thus comprises the powerset of its base type T_0.

EXAMPLES

```
TYPE BITSET =      SET OF [0 .. 15]
TYPE TapeStatus =SET OF exception  (see 1.3)
```

Given the variables

```
b: BITSET
t: ARRAY [1 .. 6] OF TapeStatus
```

particular set-structured values may be constructed and assigned, for example, as follows:

```
b := {2, 3, 5, 7, 11, 13}
t[3] := TapeStatus{manual}
t[5] := TapeStatus{}
t[6] := TapeStatus{unloaded .. skew}
```

Here, the value assigned to t_3 is the *singleton set* consisting of the single element *manual*; to t_5 is assigned the empty set, meaning that the fifth tape unit is returned to operational (non-exceptional) status, whereas t_6 is assigned the set of all three exceptions.

The cardinality of a set type T is

$$card(T) = 2^{card(T0)} \quad (1.17)$$

This can easily be derived from the fact that each of the card(T_0) elements of T_0 must be represented by one of the two values *present* or *absent* and that all elements are independent of each other. It is evidently essential for an efficient and economical implementation that the base type be not only finite, but that its cardinality is reasonably small.

The following elementary operators are defined on all set types:

*	set intersection
+	set union
-	set difference
IN	set membership

Constructing the *intersection* or the *union* of two sets is often called *set multiplication* or *set addition,* respectively; the priorities of the set operators are defined accordingly, with the intersection operator having priority over the union and difference operators, which in turn have priority over the membership operator, which is classified as a relational operator. Following are examples of set expressions and their fully parenthesized equivalents:

```
r * s + t      = (r*s) + t
r - s * t      = r - (s*t)
r - s + t      = (r-s) + t
x IN s + t     = x IN (s+t)
```

1.10. REPRESENTATION OF ARRAY, RECORD, AND SET STRUCTURES

The essence of the use of abstractions in programming is that a program may be conceived, understood, and verified on the basis of the laws governing the abstractions, and that it is not necessary to have further insight and knowledge about the ways in which the abstractions are implemented and represented in a particular computer. Nevertheless, it is essential for a professional programmer to have an understanding of widely used techniques for representing the basic concepts of programming abstractions, such as the fundamental data structures. It is helpful insofar as it might enable the programmer to make sensible decisions about program and data design in the light not only of the abstract properties of structures, but also of their realizations on actual computers, taking into account a computer's particular capabilities and limitations.

The problem of data representation is that of mapping the abstract structure into a computer store. Computer stores are - in a first approximation - arrays of individual storage cells called *words.* The indices of the words are called *addresses.*

```
VAR store: ARRAY ADDRESS OF WORD
```

The cardinalities of the types ADDRESS and WORD vary from one computer to another. A particular problem is the great variability of the cardinality of the word. Its logarithm is called the *wordsize,* because it is the number of bits that a storage cell consists of.

1.10.1. Representation of Arrays

A representation of an array structure is a mapping of the (abstract) array with components of type T onto the store which is an array with components of type WORD. The array should be mapped in such a way that the computation of addresses of array components is as simple (and therefore as efficient) as possible. The address i of the j-th array component is computed by the linear mapping function

$$i = i_0 + j*s \tag{1.18}$$

where i_0 is the address of the first component, and s is the number of words that a component occupies. Assuming that the word is the smallest individually accessible unit of store, it is evidently highly desirable that s be a whole number, the simplest case being $s = 1$. If s is not a whole number (and this is the normal case), then s is usually rounded up to the next larger integer $\lceil s \rceil$. Each array component then occupies $\lceil s \rceil$ words, whereby $\lceil s \rceil - s$ words are left unused (see Figs. 1.5 and 1.6). Rounding up of the number of words needed to the next whole number is called *padding.* The storage utilization factor u is the quotient of the minimal amounts of storage needed to represent a structure and of the amount actually used:

$$u = s / \lceil s \rceil \tag{1.19}$$

Since an implementor has to aim for a storage utilization as close to 1 as possible, and since accessing parts of words is a cumbersome and relatively inefficient process, he or she

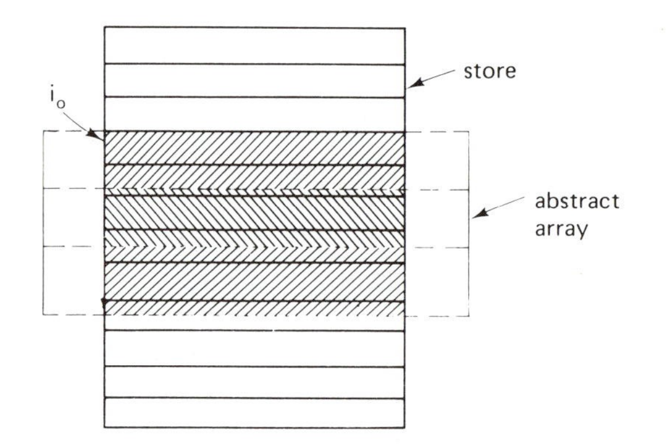

Fig. 1.5 Mapping an array onto a store.

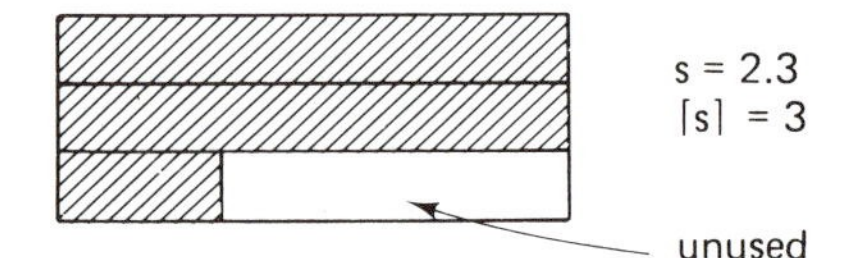

Fig. 1.6 Padded representation of a record.

must compromise. The following considerations are relevant:

1. Padding decreases storage utilization.
2. Omission of padding may necessitate inefficient partial word access.
3. Partial word access may cause the code (compiled program) to expand and therefore to counteract the gain obtained by omission of padding.

In fact, considerations 2 and 3 are usually so dominant that compilers always use padding automatically. We notice that the utilization factor is always u > 0.5, if s > 0.5. However, if s ≤ 0.5, the utilization factor may be significantly increased by putting more than one array component into each word. This technique is called *packing.* If n components are packed into a word, the utilization factor is (see Fig. 1.7)

$$u = n*s / \lceil n*s \rceil \tag{1.20}$$

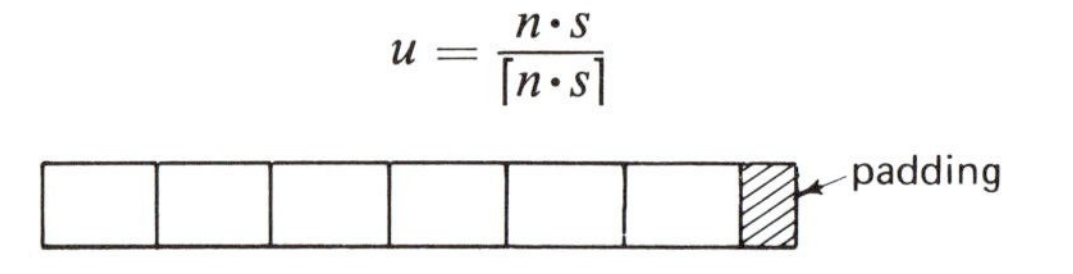

Fig. 1.7 Packing six components into one word.

Access to the i-th component of a packed array involves the computation of the word address j in which the desired component is located, and it involves the computation of the respective component position k within the word.

$$j = i \text{ DIV } n \qquad k = i \text{ MOD } n$$

In most programming languages the programmer is given no control over the representation of the abstract data structures. However, it should be possible to indicate the desirability of packing at least in those cases in which more than one component would fit into a single word, i.e., when a gain of storage economy by a factor of 2 and more could be achieved. We propose the convention to indicate the desirability of packing by prefixing the symbol ARRAY (or RECORD) in the declaration by the symbol PACKED.

1.10.2. Representation of Record Structures

Records are mapped onto a computer store by simply juxtaposing their components. The address of a component (field) r_i relative to the origin address of the record r is called the field's *offset* k_i. It is computed as

$$k_i = s_1 + s_2 + ... + s_{i-1} \tag{1.21}$$

where s_j is the size (in words) of the j-th component. We now realize that the fact that all components of an array are of equal type has the welcome consequence that $k_i = (i-1)*s$. The generality of the record structure does unfortunately not allow such a simple, linear function for offset address computation, and it is therefore the very reason for the requirement that record components be selectable only by fixed identifiers. This restriction has the desirable benefit that the respective offsets are known at compile time. The resulting greater efficiency of record field access is well-known.

The technique of packing may be beneficial, if several record components can be fitted into a single storage word (see Fig. 1.8). Since offsets are computable by the compiler, the offset of a field packed within a word may also be determined by the compiler. This means that on many computers packing of records causes a deterioration in access efficiency considerably smaller than that caused by the packing of arrays.

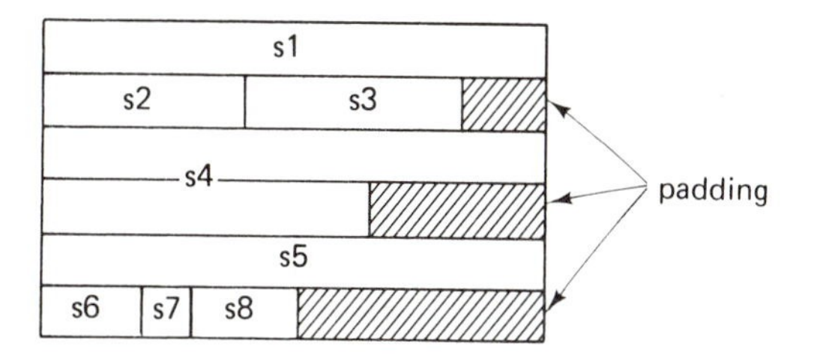

Fig. 1.8 Representation of a packed record.

1.10.3. Representation of Sets

A set s is conveniently represented in a computer store by its *characteristic function* C(s).

This is an array of logical values whose ith component has the meaning *i is present in s*. The size of the array is determined by the set type's cardinality.

$$C(s_i) = (i \text{ IN } s) \tag{1.22}$$

As an example, the set of small integers s = {2, 3, 5, 7, 11, 13} is represented by the sequence of logical values F(alse) and T(rue)

$$C(s) = (FFTTFTFTFFFTFTFF)$$

if the base type of s is the integer subrange 0 .. 15. In a computer store the sequence of logical values is represented as a so-called *bitstring*, i.e. a sequence of zeroes and ones.

The representation of sets by their characteristic function has the advantage that the operations of computing the union, intersection, and difference of two sets may be implemented as elementary logical operations. The following equivalences, which hold for all elements i of the base type of the sets x and y, relate logical operations with operations on sets:

$$\begin{aligned} i \text{ IN } (x+y) &= (i \text{ IN } x) \text{ OR } (i \text{ IN } y) \\ i \text{ IN } (x*y) &= (i \text{ IN } x) \text{ AND } (i \text{ IN } y) \\ i \text{ IN } (x-y) &= (i \text{ IN } x) \text{ AND NOT } (i \text{ IN } y) \end{aligned} \tag{1.23}$$

These logical operations are available on all digital computers, and moreover they operate concurrently on all corresponding elements (bits) of a word. It therefore appears that in order to be able to implement the basic set operations in an efficient manner, sets must be represented in a small, fixed number of words upon which not only the basic logical operations, but also those of shifting are available. Testing for membership is then implemented by a single shift and a subsequent (sign) bit test operation. As a consequence, a test of the form x IN $\{c_1, c_2, \dots, c_n\}$ can be implemented considerably more efficiently than the equivalent Boolean expression

$$(x = c_1) \text{ OR } (x = c_2) \text{ OR } \dots \text{ OR } (x = c_n)$$

A corollary is that the set structure should be used only in the case of small base types. The limit of the cardinality of base types for which a reasonably efficient implementation can be guaranteed is detemined by the wordlength of the underlying computer, and it is plain that computers with large wordlengths are preferred in this respect. If the wordsize is relatively small, a representation using multiple words for a set may be chosen.

1.11. THE SEQUENCE STRUCTURE

The fourth elementary structuring method is the *sequence.* A sequence type might be declared as

$$\text{TYPE T} = \text{SEQUENCE OF } T_0 \tag{1.24}$$

This form makes clear that all elements in the sequence are of the same type, called the *base type* T_0 of the sequence. We shall denote a sequence s with n elements by

$$s = \langle s_0, s_1, s_2, \dots, s_{n-1} \rangle$$

n is called the *length* of the sequence. This structure looks quite similar to the array. The essential difference is that in the case of the array the number of elements is fixed by the array's declaration, whereas for the sequence it is left open. This implies that it may vary during execution of the program. Although every sequence has at any time a specific, finite length, we must consider the cardinality of a sequence type as infinite, because there is no fixed limit to the potential length of sequence variables.

A direct consequence of the infinite cardinality of sequence types is the impossibility to allocate a given amount of storage to sequence variables. Instead, storage has to be allocated during program execution, namely whenever the sequence grows. Perhaps storage can be reclaimed when the sequence shrinks. In any case, a dynamic storage allocation scheme must be employed. All structures with infinite cardinality share this property, which is so essential that we classify them as *advanced structures* in contrast to the fundamental structures discussed so far.

What, then, causes us to place the discussion of sequences in this chapter on fundamental structures? The primary reason is that the storage management strategy is sufficiently simple for sequences (in contrast to other advanced structures), if we enforce a certain discipline in the use of sequences. In fact, under this proviso the handling of storage can safely be delegated to a machanism that can be guaranteed to be reasonably effective. The secondary reason is that sequences are indeed ubiquitous in all computer applications. This structure is prevalent in all cases where different kinds of storage media are involved, i.e. where data are to be moved from one medium to another, such as from disk or tape to primary store or vice-versa.

The discipline mentioned is the restraint to use *sequential access* only. By this we mean that a sequence is inspected by strictly proceeding from one element to its immediate successor, and that it is generated by repeatedly appending an element at its end. The immediate consequence is that elements are not directly accessible, with the exception of the one element which currently is up for inspection. It is this accessing discipline which fundamentally distinguishes sequences from arrays. As we shall see in Chapter 2, the influence of an access discipline on programs is profound.

The advantage of adhering to sequential access which, after all, is a serious restriction, is the relative simplicity of needed storage management. But even more important is the possibility to use effective buffering techniques when moving data to or from secondary storage devices. Sequential access allows us to feed streams of data through pipes between the different media. Buffering implies the collection of sections of a stream in a so-called *buffer,* and the subsequent shipment of the whole buffer content once the buffer is filled. This results in very significantly more effective use of secondary storage. Given sequential access only, the buffering mechanism is reasonably straightforward for all sequences and all media. It can therefore safely be built into a system for general use, and the programmer need not be burdened by incorporating it in his program. Such a system is usually called a *filing system,* because the high-volume, sequential access devices are used for permanent storage of data, and they retain them even when the computer is switched off. The unit of data on these media is the sequence or *sequential file.*

There exist certain storage media in which the sequential access is indeed the only possible one. Among them are evidently all kinds of tapes. But even on magnetic disks each recording track constitutes a storage facility allowing only sequential access. Strictly sequential access is the primary characteristic of every mechanically moving device and of some other ones as well.

We summarize the essence of the foregoing as follows:

1. Arrays, records, and sets are random access structures. They are used when located in primary, random-access store.

2. Sequences are used to access data on secondary, sequential-access stores, such as disks and tapes.

1.11.1 Elementary Sequence Operators

The discipline of sequential access can be enforced by providing a set of seqencing operators through which sequence variables can be accessed *exclusively.* Hence, although we may here refer to the i-th element of a sequence s by writing s_i, this shall not be possible in a program. Evidently, the set of operators must contain an operator for generating and one for inspecting a sequence. As already mentioned, a sequence is generated by appending elements at its end, and it is inspected by proceeding to the next element. Hence, a certain *position* is always associated with every sequence. We shall now postulate and describe informally the following set of primitive operators:

1. Open(s) defines s to be the empty sequence, i.e. the sequence of length 0.

2. Write(s, x) appends an element with value x to the sequence s.

3. Reset(s) sets the current position to the beginning of s.

4. Read(s, x) assigns the element designated by the current position to the variable x and advances the position to the next element.

In order to convey a more precise understanding of the sequencing operators, the following example is provided. It shows how they *might* be expressed when sequences are represented in terms of arrays. This example of an implementation intentionally builds upon concepts introduced and discussed earlier, and it does not involve either buffering or sequential stores which, as mentioned above, make the sequence concept truly attractive.

Nevertheless, this example exhibits all the essential characteristics of the primitive sequence operators when they represent a sequential file system for backing stores.

First, the operators are presented in terms of conventional procedures. This collection of definitions is called a *definition module*. Since the type of the sequence appears in the formal parameter lists, this type must also be declared. The declaration is a good example of an application of a record structure because, in addition to the field denoting the array which represents the sequence, further fields are required to denote the current length and position, i.e. the *state* of the sequence. Furthermore, we add another field which shall indicate whether a read operation had been successful, or whether it failed because there was no next element to be read.

```
DEFINITION MODULE FileSystem;                                    (1.25)
 FROM SYSTEM IMPORT WORD;

 CONST MaxLength = 4096;
 TYPE Sequence = RECORD pos, length: CARDINAL;
                   eof: BOOLEAN;
                   a: ARRAY [0 .. MaxLength-1] OF WORD
                 END ;

 PROCEDURE Open(VAR f: Sequence);
 PROCEDURE WriteWord(VAR f: Sequence; w: WORD);
 PROCEDURE Reset(VAR f: Sequence);
 PROCEDURE ReadWord(VAR f: Sequence; VAR w: WORD);
 PROCEDURE Close(VAR f: Sequence);
END FileSystem.
```

Note that in this example the maximum length that sequences may reach is an arbitrary constant. Should a program cause a sequence to become longer, then this would not be a mistake of the program, but an inadequacy of this implementation. On the other hand, a read operation proceeding beyond the current end of the sequence would indeed be the program's mistake. In order to avoid it, a facility for inspecting whether the end is reached must be provided. It is present in the form of the Boolean field *eof* (for *e*nd *of*). It is the only field that is supposed to be known to the user of this implementation of the abstraction of the sequence type. The base type is here chosen to be WORD. It shall stand for any type in general. In the language Modula-2 it is a type that is parameter-compatible with several standard types, including INTEGER.

The statements that constitute the procedures are sepecified in the corresponding implementation module:

```
IMPLEMENTATION MODULE FileSystem;
 FROM SYSTEM IMPORT WORD;                                        (1.26)

 PROCEDURE Open(VAR f: Sequence);
 BEGIN f.length := 0; f.pos := 0; f.eof := FALSE
 END Open;

 PROCEDURE WriteWord(VAR f: Sequence; w: WORD);
 BEGIN
  WITH f DO
```

```
      IF pos < MaxLength THEN
        a[pos] := w; pos := pos+1; length := pos
      ELSE HALT
      END
    END
  END WriteWord;

  PROCEDURE Reset(VAR f: Sequence);
  BEGIN f.pos := 0; f.eof := FALSE
  END Reset;

  PROCEDURE ReadWord(VAR f: Sequence; VAR w: WORD);
  BEGIN
    WITH f DO
      IF pos = length THEN f.eof := TRUE
        ELSE w := a[pos]; pos := pos+1
      END
    END
  END ReadWord;

  PROCEDURE Close(VAR f: Sequence);
  BEGIN (*empty*)
  END Close;
END FileSystem.
```

Given this set of sequence operators, the following program schemata emerge for writing and subsequent reading of a sequential file s:

```
Open(s);
WHILE B DO P(x); WriteWord(s, x) END                          (1.27)

Reset(s); ReadWord(s, x);
WHILE ~s.eof DO Q(x); ReadWord(s, x) END
```

Here B is the condition that must hold before the statement P is executed to compute the value of a next element, which is then appended to the sequence s. When reading, Q is the statement that is applied to each value read from the sequence. Q is guarded by the condition *~s.eof* which guarantees that indeed a next element had been read. We note a slight asymmetry between the two schemata. The additional read statement is due to the fact that for reading n elements n+1 read operations must be executed, of which the last one fails. This unsuccessful read statement is needed, because the read operator is the only one which changes the value of the variable *eof*. It could be avoided by postulating that the reset operator would define the value of the field *eof* as *length = 0*. The language Pascal, for example, defines the reset operator in this sensible way. But most widely used file systems do not, and this is why we adhere to this strategy in demonstrating the use of sequences.

Quite frequently, file systems allow some relaxation of the discipline of strictly sequential access without grave penalty. In particular, they allow the positioning of the sequence at an arbitrary place instead of the beginning only. This relaxation of rules is accomplished by the introduction of an additional sequence procedure called *SetPos(s, k)* with $0 \leq k < s.length$. Evidently, *SetPos(s, 0)* is equivalent to *Reset(s)*. The ease of implementing this extension is due to our definition of the writing operation, which postulates that writing always occur at the sequence's end. Consequently, if a write operation occurs when the file position is not at the end, the new element replaces the entire previous tail, i.e. the sequence is truncated.

1.11.2. Buffering sequences

When data are transferred to or from a secondary storage device, the individual bits are transferred as a stream. Usually, a device imposes strict timing constraints upon the transmission. For example, if data are written on a tape, the tape moves at a fixed speed and requires the data to be fed at a fixed rate. When the source ceases, the tape movement is switched off and speed decreases quickly, but not instantaneously. Thus a gap is left between the data transmitted and the data to follow at a later time. In order to achieve a high density of data, the number of gaps ought to be kept small, and therefore data are transmitted in relatively large *blocks* once the tape is moving. Similar conditions hold for magnetic disks, where the data are allocated on tracks with a fixed number of blocks of fixed size, the so-called *block size.* In fact, a disk should be regarded as an array of blocks, each block being read or written as a whole, containing typically 2^k bytes with $k = 8, 9$, or 10.

Our programs, however, do not observe any such timing constraints. In order to allow them to ignore the constraints, the data to be transferred are *buffered.* They are collected in a so-called *buffer variable* (in main store) and transferred when a sufficient amount of data is accumulated to form a block of the required size.

Buffering has an additional advantage in allowing the process which generates (receives) data to proceed concurrently with the device that writes (reads) the data from (to) the buffer. In fact, it is convenient to regard the device as a process itself which merely copies data streams. The buffer's purpose is to provide a certain degree of decoupling between the two processes, which we shall call the *producer* and the *consumer.* If, for example, the consumer is slow at a certain moment, it may catch up with the producer later on. This decoupling is often essential for a good utilization of peripheral devices, but it has only an effect, if the rates of producer and consumer are about the same on the average, but fluctuate at times. The degree of decoupling increases with increasing buffer size.

We now turn to the question of how to represent a buffer, and shall for the time being assume that data elements are deposited and fetched individually instead of in blocks. A buffer essentially constitutes a first-in-first-out queue. If it is declared as an array, two index variables, say *in* and *out,* mark the positions of the next location to be written into and to be read from. Ideally, such an array should have no index bounds. A finite array is quite adequate, however, considering the fact that elements once fetched are no longer relevant. Their location may well be re-used. This leads to the idea of the *circular buffer.* The operations of depositing and fetching an element are expressed in the following module, which exports these operations as procedures, but hides the buffer and its index variables - and thereby effectively the buffering mechanism - from the client processes. This mechanism also involves a variable *n* counting the number of elements currently in the buffer. If N denotes the size of the buffer, the condition $0 \leq n \leq N$ is an obvious invariant. Therefore, the operation *fetch* must be guarded by the condition $n > 0$ (buffer non-empty), and the operation *deposit* by the condition $n < N$ (buffer non-full). Not meeting the former condition must be regarded as a programming error, a violation of the latter as a failure of the suggested implementation (buffer too small).

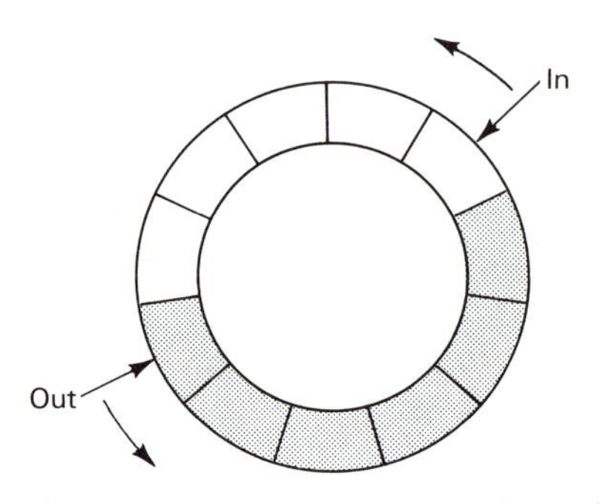

Fig. 1.9 Circular buffer with indices *in* and *out*.

```
MODULE Buffer;
  EXPORT deposit, fetch;                                          (1.28)
  CONST N = 1024; (*buffer size*)
  VAR n, in, out: CARDINAL;
    buf: ARRAY [0 .. N-1] OF WORD;
  PROCEDURE deposit(x: WORD);
  BEGIN
    IF n = N THEN HALT END ;
    n := n+1; buf[in] := x; in := (in + 1) MOD N
  END deposit;
  PROCEDURE fetch(VAR x: WORD);
  BEGIN
    IF n = 0 THEN HALT END ;
    n := n-1; x := buf[out]; out := (out + 1) MOD N
  END fetch;
BEGIN n := 0; in := 0; out := 0
END Buffer.
```

This implementation of a buffer is acceptable only, if the procedures *deposit* and *fetch* are activated by a single agent (once acting as a producer, once as a consumer). If, however, they are activated by individual, concurrent processes, this scheme is too simplistic. The reason is that the attempt to deposit into a full buffer, or the attempt to fetch from an empty buffer, are quite legitimate. The execution of these actions will merely have to be delayed until the guarding conditions are established. Such delays essentially constitute the necessary *synchronization* among concurrent processes. We may represent these delays respectively by the statements

```
REPEAT UNTIL n < N
REPEAT UNTIL n > 0
```

which must be substituted for the two conditioned HALT statements in (1.28). This solution is, however, not recommended, even if it known that the two processes are driven by two individual engines. The reason is that the two processors necessarily access the same variable n, and therefore the same store. The idling process, by constantly polling the value n, hinders its partner, because at no time can the store be accessed by more than one process. This kind of *busy waiting* must indeed be avoided, and we therefore postulate a facility that makes the details of synchronization less explicit, in fact hides them. We shall call this facility a *signal*, and assume that it is available from a utility module (called *Processes*) together with a set of primitive operators on signals.

Every signal s is associated with a guard (condition) P_s. If a process needs to be delayed until P_s is established (by some other process), it must, before proceeding, wait for the signal s. This is to be expressed by the statement *Wait(s)*. If, on the other hand, a process establishes P_s, it thereupon signals this fact by the statement *Send(s)*. If P_s is an established precondition to every statement *Send(s)*, then P_s can be regarded as a postcondition of *Wait(s)*.

```
DEFINITION MODULE Processes;
  TYPE Signal;                                        (1.29)
  PROCEDURE Wait(VAR s: Signal);
  PROCEDURE Send(VAR s: Signal);
  PROCEDURE Init(VAR s: Signal);
END Processes.
```

We are now able to express the buffer module (1.28) in a form that functions properly when used by individual, concurrent processes:

```
MODULE Buffer;
  IMPORT Signal, Wait, Send, Init;                    (1.30)
  EXPORT deposit, fetch;
  CONST N = 1024; (*buffer size*)
  VAR n, in, out: CARDINAL;
    nonfull: Signal;  (*n < N*)
    nonempty: Signal; (*n > 0*)
    buf: ARRAY [0 .. N-1] OF WORD;
  PROCEDURE deposit(x: WORD);
  BEGIN
    IF n = N THEN Wait(nonfull) END ;
    n := n+1; buf[in] := x; in := (in + 1) MOD N;
    IF n = 1 THEN Send(nonempty) END
  END deposit;
  PROCEDURE fetch(VAR x: WORD);
  BEGIN
    IF n = 0 THEN Wait(nonempty) END ;
    n := n-1; x := buf[out]; out := (out + 1) MOD N;
    IF n = N-1 THEN Send(nonfull) END
  END fetch;
```

```
BEGIN n := 0; in := 0; out := 0; Init(nonfull); Init(nonempty)
END Buffer.
```

An additional caveat must be made, however. The scheme fails miserably, if by coincidence both consumer and producer fetch the counter value n simultaneously for updating. Unpredictably, its resulting value will be either n+1 or n-1, but not n. It is indeed necessary to protect the processes from dangerous interference. In general, all operations that *alter the values of shared variables* constitute potential pitfalls.

A sufficient (but not always necessary) condition is that all shared variables be declared local to a module whose procedures are guaranteed to be executed under *mutual exclusion*. Such a module is called a *monitor* [1-7]. The mutual exclusion provision guarantees that at any time at most one process is actively engaged in executing a procedure of the monitor. Should another process be calling a procedure of the (same) monitor, it will automatically be delayed until the first process has terminated its procedure.

Note: By actively engaged is meant that a process execute a statement other than a wait statement.

At last we return now to the problem where the producer or the consumer (or both) require the data to be available in a certain block size. The following module is a variant of (1.30), assuming a block size of N_p data elements for the producer, and of N_c elements for the consumer. In these cases, the buffer size N is usually chosen as a common multiple of N_p and N_c. In order to emphasise that symmetry between the operations of fetching and depositing data, the single counter *n* is now represented by two counters, namely *ne* and *nf*. They specify the numbers of empty and filled buffer slots respectively. When the consumer is idle, *nf* indicates the number of elements needed for the consumer to proceed; and when the producer is waiting, *ne* specifies the number of elements needed for the producer to resume. (Therefore ne+nf = N does not always hold).

```
MODULE Buffer;
  IMPORT Signal, Wait, Send, Init;                                    (1.31)
  EXPORT deposit, fetch;
  CONST Np = 16;  (*size of producer block*)
    Nc = 128;  (*size of consumer block*)
    N = 1024;  (*buffer size, common multiple of Np and Nc*)
  VAR ne, nf: INTEGER;
    in, out: CARDINAL;
    nonfull: Signal;   (*ne >= 0*)
    nonempty: Signal;  (*nf >= 0*)
    buf: ARRAY [0 .. N-1] OF WORD;
  PROCEDURE deposit(VAR x: ARRAY OF WORD);
  BEGIN ne := ne - Np;
    IF ne < 0 THEN Wait(nonfull) END ;
    FOR i := 0 TO Np-1 DO buf[in] := x[i]; in := in + 1 END ;
    IF in = N THEN in := 0 END ;
    nf := nf + Np;
    IF nf >= 0 THEN Send(nonempty) END
  END deposit;
```

```
  PROCEDURE fetch(VAR x: ARRAY OF WORD);
  BEGIN nf := nf - Nc;
    IF nf < 0 THEN Wait(nonempty) END ;
    FOR i := 0 TO Nc-1 DO x[i] := buf[out]; out := out + 1 END;
    IF out = N THEN out := 0 END ;
    ne := ne + Nc;
    IF ne >= 0 THEN Send(nonfull) END
  END fetch;
BEGIN
  ne := N; nf := 0; in := 0; out := 0; Init(nonfull); Init(nonempty)
END Buffer.
```

1.11.3 Standard Input and Output

By standard input and output we understand the transfer of data to (from) a computer system from (to) genuinely external agents, in particular its human operator. Input may typically originate at a keyboard and output may sink into a display screen. In any case, its characteristic is that it is readable, and it typically consists of a sequence of characters. This readability condition is responsible for yet another complication incurred in most genuine input and output operations. Apart from the actual data transfer, they also involve a *transformation of representation*. For example, numbers, usually considered as atomic units and represented in binary form, need be transformed into readable, decimal notation. Structures need to be represented in a suitable layout, whose generation is called *formatting*.

Whatever the transformation may be, the concept of the sequence is once again instrumental for a considerable simplification of the task. The key is the observation that, if the data set can be considered as a sequence, the transformation of the sequence can be implemented as a sequence of (identical) transformations of elements.

$$T(\langle s_0, s_1, \dots, s_{n-1}\rangle) = \langle T(s_0), T(s_1), \dots, T(s_{n-1})\rangle \qquad (1.32)$$

We shall briefly investigate the necessary operations for transforming representations of natural numbers for input and output. The basis is that a number x represented by the sequence of decimal digits $d = \langle d_{n-1}, \dots, d_1, d_0\rangle$ has the value

$$x = \mathbf{S}i: 0 \le i < n : d_i * 10^i$$

Assume now that the sequence d is to be read and transformed, and the resulting numeric value to be assigned to x. The simple algorithm is given in (1.33); it terminates with the reading of the first character that is not a digit. (Arithmetic overflow is not considered).

```
x := 0; Read(ch);
WHILE ("0" <= ch) & (ch <= "9") DO                                    (1.33)
  x := 10*x + (ORD(ch) - ORD("0")); Read(ch)
END
```

In the case of output the transformation is complexified by the fact that the decomposition of x into decimal digits yields them in the reverse order. The least digit is generated first by computing x MOD 10. This requires an intermediate buffer in the form of a first-in-last-out queue (stack). We represent it as an array d with index i and obtain the following program:

```
i := 0;
REPEAT d[i] := x MOD 10; x := x DIV 10; i := i+1                    (1.34)
UNTIL x = 0;
REPEAT i := i-1; Write(CHR(d[i] + ORD("0"))
UNTIL i = 0
```

Note: A consistent substitution of the constant 10 in (1.33) and (1.34) by a constant B (>0) will yield number conversion routines to and from representations with base B. Frequently used cases are B = 8 (octal) and B = 16 (hexadecimal), because the involved multiplications and divisions can be implemented by simple shifts of the binary numbers.

Obviously, it should not be necessary to specify these ubiquitous operations in every program in full detail. We therefore postulate a utility module that provides the most common standard input and output operations on numbers and strings. This module is referenced in most programs throughout this book, and we call it *InOut*. Its procedures serve for reading and writing a character, an integer, a cardinal number, or a string. Because of its frequent use in this book, the module's definition part is here listed in full.

```
DEFINITION MODULE InOut;
 FROM SYSTEM IMPORT WORD;                                            (1.35)
 FROM FileSystem IMPORT File;

 CONST EOL = 36C;
 VAR Done: BOOLEAN;
  termCH: CHAR; (*terminating character in ReadInt, ReadCard*)
  in, out: File; (*for exceptional cases only*)

 PROCEDURE OpenInput(defext: ARRAY OF CHAR);
  (*request a file name and open input file "in".
   Done := "file was successfully opened".
   If open, subsequent input is read from this file.
   If name ends with ".", append extension defext*)

 PROCEDURE OpenOutput(defext: ARRAY OF CHAR);
  (*request a file name and open output file "out"
   Done := "file was successfully opened.
   If open, subsequent output is written on this file*)

 PROCEDURE CloseInput;
  (*closes input file; returns input to terminal*)

 PROCEDURE CloseOutput;
  (*closes output file; returns output to terminal*)

 PROCEDURE Read(VAR ch: CHAR);
  (*Done := NOT in.eof*)

 PROCEDURE ReadString(VAR s: ARRAY OF CHAR);
  (*read string, i.e. sequence of characters not containing
   blanks nor control characters; leading blanks are ignored.
```

```
   Input is terminated by any character <= " ";
   this character is assigned to termCH.
   DEL is used for backspacing when input from terminal*)

 PROCEDURE ReadInt(VAR x: INTEGER);
  (*read string and convert to integer. Syntax:
    integer = ["+"|"-"] digit {digit}.
   Leading blanks are ignored.
   Done := "integer was read"*)

 PROCEDURE ReadCard(VAR x: CARDINAL);
  (*read string and convert to cardinal. Syntax:
    cardinal = digit {digit}.
   Leading blanks are ignored.
   Done := "cardinal was read"*)

 PROCEDURE ReadWrd(VAR w: WORD);
  (*Done := NOT in.eof*)

 PROCEDURE Write(ch: CHAR);

 PROCEDURE WriteLn;   (*terminate line*)

 PROCEDURE WriteString(s: ARRAY OF CHAR);

 PROCEDURE WriteInt(x: INTEGER; n: CARDINAL);
  (*write integer x with (at least) n characters on file "out".
   If n is greater than the number of digits needed,
   blanks are added preceding the number*)

 PROCEDURE WriteCard(x, n: CARDINAL);
 PROCEDURE WriteOct(x, n: CARDINAL);
 PROCEDURE WriteHex(x, n: CARDINAL);
 PROCEDURE WriteWrd(w: WORD);
END InOut.
```

This module's read and write procedures do not have a parameter indicating the sequence (file) affected. Instead, it is implicitly understood that data are read from the standard input device (keyboard) and written to the standard ouput device. This assumption can, however, be overridden by invoking the procedures *OpenInput* and *OpenOutput*. They request a file name from the operator, and substitute the specified file for the standard device, thereby rerouting the data stream. Procedures *CloseInput* and *CloseOutput* return the reading and writing to the default devices.

1.12 SEARCHING

The task of searching is one of most frequent operations in computer programming. It also provides an ideal ground for application of the data structures so far encountered. There exist several basic variations of the theme of searching, and many different algorithms have been developed on this subject. The basic assumption in the following presentations is that the collection of data, among which a given element is to be searched, is *fixed.* We shall assume that this set of N elements is represented as an array, say as

a: ARRAY [0 .. N-1] OF item

Typically, the type *item* has a record structure with a field that acts as a key. The task then consists of finding an element of *a* whose key field is equal to a given search argument x. The resulting index i, satisfying *a[i].key = x,* then permits access to the other fields of the located element. Since we are here interested in the task of searching only, and do not care about the data for which the element was searched in the first place, we shall assume that the type *item* consists of the key only, i.e *is* the key.

1.12.1. Linear Search

When no further information is given about the searched data, the obvious approach is to proceed sequentially through the array in order to increase step by step the size of the section, where the desired element is known not to exist. This approach is called *linear search.* There are two conditions which terminate the search:

1. The element is found, i.e. $a_i = x$.
2. The entire array has been scanned, and no match was found.

This results in the following algorithm:

```
i := 0;                                                    (1.36)
WHILE (i < N) & (a[i] # x) DO i := i+1 END
```

Note that the order of the terms in the Boolean expression is relevant. The invariant, i.e the condition satisfied before each incrementing of the index i, is

$$(0 \leq i < N) \,\&\, (Ak : 0 \leq k < i : a_k \neq x) \qquad (1.37)$$

expressing that for all values of k less than i no match exists. From this and the fact that the search terminates only if the condition in the while-clause is false, the resulting condition is derived as

$$((i = N) \text{ OR } (a_i = x)) \,\&\, (Ak : 0 \leq k < i : a_k \neq x)$$

This condition not only is our desired result, but also implies that when the algorithm did find a match, it found the one with the least index, i.e. the first one. i = N implies that no match exists.

Termination of the repetition is evidently guaranteed, because in each step i is increased and therefore certainly will reach the limit N after a finite number of steps; in fact, after N steps, if no match exists.

Each step evidently requires the incrementing of the index and the evaluation of a Boolean expression. Could this task be simplifed, and could the search thereby be speeded up? The only possibility lies in finding a simplification of the Boolean expression which notably consists of two factors. Hence, the only chance for finding a simpler solution lies in establishing a condition consisting of a single factor that implies both factors. This is possible only by guaranteeing that a match will be found, and is achieved by posting an additional element with value x at the end of the array. We call this auxiliary element a *sentinel,* because it prevents the search from passing beyond the index limit. The array a is now declared as

```
a: ARRAY [0 .. N]  OF INTEGER
```

and the linear search algorithm with sentinel is expressed by

```
a[N] := x; i := 0;                                    (1.38)
WHILE a[i] # x DO i := i+1 END
```

The resulting condition, derived from the same invariant as before, is

$$(a_i = x) \& (Ak : 0 \leq k < i : a_k \neq x)$$

Evidently, i = N implies that no match (except that for the sentinel) was encountered.

1.12.2. Binary Search

There is quite obviously no way to speed up a search, unless more information is available about the searched data. It is well known that a search can be made much more effective, if the data are ordered. Imagine, for example, a telephone directory in which the names were not alphabetically listed. It would be utterly unusable. We shall therefore present an algorithm which makes use of the knowledge that a is ordered, i.e., of the condition

$$Ak: 1 \leq k < N : a_{k-1} \leq a_k \tag{1.39}$$

The key idea is to inspect an element picked at random, say a_m, and to compare it with the search argument x. If it is equal to x, the search terminates; if it is less than x, we infer that all elements with indices less or equal to m can be eliminated from further searches; and if it is greater than x, all with index greater or equal to m can be eliminated. This results in the following algorithm called *binary search*; it uses two index variables L and R marking the Left and at the Right end of the section of a in which an element may still be found.

```
L := 0; R := N-1; found := FALSE ;                    (1.40)
WHILE (L ≤ R) & ~found DO
  m := any value between L and R;
  IF a[m] = x THEN found := TRUE
  ELSIF a[m] < x THEN L := m+1
  ELSE R := m-1
```

```
    END
  END
```

The loop invariant, i.e. the condition satisfied before each step, is

$$(L \leq R) \ \& \ (Ak : 0 \leq k < L : a_k < x) \ \& \ (Ak : R < k < N : a_k > x) \tag{1.41}$$

from which the result is derived as

$$\text{found OR } ((L > R) \ \& \ (Ak : 0 \leq k < L : a_k < x) \ \& \ (Ak : R < k < N : a_k > x))$$

which implies

$$(a_m = x) \text{ OR } (Ak : 0 \leq k < N : a_k \neq x)$$

The choice of m is apparently arbitrary in the sense that correctness does not depend on it. But it does influence the algorithm's effectiveness. Clearly our goal must be to eliminate in each step as many elements as possible from further searches, no matter what the outcome of the comparison is. The optimal solution is to choose the middle element, because this eliminates half of the array in any case. As a result, the maximum number of steps is log N, rounded up to the nearest integer. Hence, this algorithm offers a drastic improvement over linear search, where the expected number of comparisons is N/2.

The efficiency can be somewhat improved by interchanging the two if-clauses. Equality should be tested second, because it occurs only once and causes termination. But more relevant is the question, whether -- as in the case of linear search -- a solution could be found that allows a simpler condition for termination. We indeed find such a faster algorithm, if we abandon the naive wish to terminate the search as soon as a match is established. This seems unwise at first glance, but on closer inspection we realize that the gain in efficiency at every step is greater than the loss incurred in comparing a few extra elements. Remember that the number of steps is at most log N. The faster solution is based on the following invariant:

$$(Ak : 0 \leq k < L : a_k < x) \ \& \ (Ak : R \leq k < N : a_k \geq x) \tag{1.42}$$

and the search is continued until the two sections span the entire array.

```
L := 0; R := N;                                              (1.43)
WHILE L < R DO
  m := (L+R) DIV 2;
  IF a[k] < x THEN L := m+1 ELSE R := m END
END
```

The terminating condition is $L \geq R$. Is it guaranteed to be reached? In order to establish this guarantee, we must show that under all circumstances the difference R-L is diminished in each step. $L < R$ holds at the beginning of each step. The arithmetic mean m then satisfies $L \leq m < R$. Hence, the difference is indeed diminished by either assigning m+1 to L (increasing L) or m to R (decreasing R), and the repetition terminates with $L = R$. However, the invariant and $L = R$ do not yet establish a match. Certainly, if $R = N$, no match exists. Otherwise we must take into consideration that the element a[R] had never been compared. Hence, an additional test for equality a[R] = x is necessary. In contrast to the first solution (1.40), this algorithm -- like linear search -- finds the matching element with the least index.

1.12.3 Table Search

A search through an array is sometimes also called a *table search,* particularly if the keys are themselves structured objects, such as arrays of numbers or characters. The latter is a frequently encountered case; the character arrays are called *strings* or words. Let us define a type *String* as

String = ARRAY [0 .. M-1] OF CHAR (1.44)

and let order on strings x and y with be defined as follows:

$$(x = y) \;=\; (\mathbf{A}j: 0 \le j < M : x_j = y_j)$$
$$(x < y) \;=\; \mathbf{E}i: 0 \le i < N : ((\mathbf{A}j: 0 \le j < i : x_j = y_j) \,\&\, (x_i < y_i))$$

In order to establish a match, we evidently must find all characters of the comparands to be equal. Such a comparison of structured operands therefore turns out to be a search for an unequal pair of comparands, i.e. a search for inequality. If no unequal pair exists, equality is established. Assuming that the length of the words be quite small, say less than 30, we shall use a linear search in the following solution.

In most practical applications, one wishes to consider strings as having a *variable length.* This is accomplished by associating a length indication with each individual string value. Using the type declared above, this length must not exceed the maximum length M. This scheme allows for sufficient flexibility for many cases, yet avoids the complexities of dynamic storage allocation. Two representations of string lengths are most commonly used:

1. The length is implicitly specified by appending a terminating character which does not otherwise occur. Usually, the non-printing value 0C is used for this purpose. (It is important for the subsequent applications that it be the *least* character in the character set).

2. The length is explicitly stored as the first element of the array, i.e. the string s has the form

   $$s = s_0, s_1, s_2, \dots, s_{N-1}$$

   where $s_1 \dots s_{N-1}$ are the actual characters of the string and s_0 = CHR(N). This solution has the advantage that the length is directly available, and the disadvantage that the maximum length is limited to the size of the character set (256).

For the subsequent search algorithm, we shall adhere to the first scheme. A string comparison then takes the form

```
i := 0;                                                    (1.45)
WHILE (x[i] = y[i]) & (x[i] # 0C) DO i := i+1 END
```

The terminating character now functions as a sentinel, the loop invariant is

$$Aj: 0 \leq j < i : x_j = y_j \neq 0C,$$

and the resulting condition is therefore

$$((x_i \neq y_i) \text{ OR } (x_i = 0C)) \& (Aj: 0 \leq j < i : x_j = y_j \neq 0C)$$

It establishes a match between x and y, provided that $x_i = y_i$, and it establishes $x < y$, if $x_i < y_i$.

We are now prepared to return to the task of table searching. It calls for a nested search, namely a search through the entries of the table, and for each entry a sequence of comparisons between components. For example, let the table T and the search argument x be defined as

```
T: ARRAY [0 .. N-1] OF String;
x: String
```

Assuming that N may be fairly large and that the table is alphabetically ordered, we shall use a binary search. Using the algorithms for binary search (1.43) and string comparison (1.45) developed above, we obtain the following program segment.

```
L := 0; R := N;
WHILE L < R DO                                                    (1.46)
  m := (L+R) DIV 2; i := 0;
  WHILE (T[m,i] = x[i]) & (x[i] # 0C) DO i := i+1 END ;
  IF T[m,i] < x[i] THEN L := m+1 ELSE R := m END
END ;
IF R < N THEN i := 0;
  WHILE (T[R,i] = x[i]) & (x[i] # 0C) DO i := i+1 END
END
(* (R < N) & (T[R,i] = x[i]) establish a match*)
```

1.12.4. Straight String Search

A frequently encountered kind of search is the so-called *string search.* It is characterized as follows. Given an array s of N elements and an array p of M elements, where $0 < M \leq N$, declared as

```
s: ARRAY [0 .. N-1] OF item
p: ARRAY [0 .. M-1] OF item
```

string search is the task of finding the first occurrence of p in s. Typically, the items are characters; then s may be regarded as a text and p as a pattern or word, and we wish to find the first occurrence of the word in the text. This operation is basic to every text processing system, and there is obvious interest in finding an efficient algorithm for this task. Before paying particular attention to efficiency, however, let us first present a straightforward searching algorithm. We shall call it *straight string search.*

A more precise formulation of the desired result of a search is indispensible before we attempt to specify an algorithm to compute it. Let the result be the index i which points to the first occurrence of a match of the pattern with the starting. To this end, we introduce a predicate P(i,j)

$$P(i,j) = \text{A}k : 0 \leq k < j : s_{i+k} = p_k \tag{1.47}$$

Then evidently our resulting index i must satisfy P(i,M). But this condition is not sufficient. Because the search is to locate the *first* occurrence of the pattern, P(k,M) must be false for all k < i. We denote this condition by Q(i).

$$Q(i) = \text{A}k : 0 \leq k < i : \sim P(k, M) \tag{1.48}$$

The posed problem immediately suggests to formulate the search as an iteration of comparisons, and we proposed the following approach:

```
i := -1;
REPEAT i := i+1; (* Q(i) *)
  found := P(i,M)
UNTIL found OR (i = N-M)
```

The computation of P again results naturally in an iteration of individual character comparisons. When we apply DeMorgan's theorem to P, it appears that the iteration must be a search for inequality among corresponding pattern and string characters.

$$P(i,j) = (\text{A}k : 0 \leq k < j : s_{i+k} = p_k) = (\sim\text{E}k : 0 \leq k < j : s_{i+k} \neq p_k)$$

The result of the next refinement is a repetition within a repetition. The predicates P and Q are inserted at appropriate places in the program as comments. They act as invariants of the iteration loops.

```
i := -1;                                                              (1.49)
REPEAT i := i+1; j := 0; (* Q(i) *)
  WHILE (j < M) & (s[i+j] = p[j]) DO (* P(i,j+1) *) j := j+1 END
  (* Q(i) & P(i,j) & ((j = M) OR (s[i+j] # p[j])) *)
UNTIL (j = M) OR (i = N-M)
```

The term j = M in the terminating condition indeed corresponds to the condition *found,* because it implies P(i,M). The term i = N-M implies Q(N-M) and thereby the non-existence of a match anywhere in the string. If the iteration continues with j < M, then it must do so with $s_{i+j} \neq p_j$. This implies, according to (1.47), ~P(i,j), which according to (1.48) implies Q(i+1), which establishes Q(i) after the next incrementing of i.

Analysis of straight string search. This algorithm operates quite effectively, if we can assume that a mismatch between character pairs occurs after at most a few comparisons in the inner loop. This is likely to be the case, if the cardinality of the item type is large. For text searches with a character set size of 128 we may well assume that a mismatch occurs after inspecting 1 or 2 characters only. Nevertheless, the worst case performance is rather alarming. Consider, for example, that the string consist of N-1 A's followed by a single B, and that the pattern consist of M-1 A's followed by a B. Then in the order of N*M comparisons are necessary to find the match at the end of the string. As we shall subsequently see, there fortunately exist methods that drastically improve this worst case behaviour.

1.12.5. The Knuth-Morris-Pratt String Search

Around 1970, D.E. Knuth, J.H. Morris, and V.R. Pratt invented an algorithm that requires essentially in the order of N character comparisons only, even in the worst case [1-8]. The new algorithm is based on the observation that by starting the next pattern comparison at its beginning each time, we may be discarding valuable information. After a partial match of the beginning of the pattern with corresponding characters in the string, we indeed know the last part of the string, and perhaps could have precompiled some data (from the pattern) which could be used for a more rapid advance in the text string. The following example of a search for the word *Hooligan* illustrates the principle of the algorithm. Underlined characters are those which were compared. Note that each time two compared characters do not match, the pattern is shifted all the way, because a smaller shift could not possibly lead to a full match.

```
Hoola-Hoola girls like Hooligans.
Hooligan
    Hooligan
     Hooligan
      Hooligan
          Hooligan
           Hooligan
            ......
                       Hooligan
```

Using the predicates P and Q defined by (1.47) and (1.48), the KMP-algorithm is the following:

```
i := 0; j := 0;
WHILE (j < M) & (i < N) DO                                   (1.50)
  (* Q(i-j) & P(i-j, j) *)
  WHILE (j >= 0) & (s[i] # p[j]) DO j := D END ;
  i := i+1; j := j+1
END
```

This formulation is admittedly not quite complete, because it contains an unspecified shift value D. We shall return to it shortly, but first point out that the conditions Q(i-j) and P(i-j, j) are maintained as global invariants, to which we may add the relations $0 \leq i < N$ and $0 \leq j < M$. This suggests that we must abandon the notion that i always marks the current position of the first pattern character in the text. Rather, the alignment position of the pattern is now i-j.

If the algorithm terminates due to $j = M$, the term P(i-j, j) of the invariant implies P(i-M, M), that is, according to (1.47) a match at position i-M. Otherwise it terminates with $i = N$, and since $j < M$, the invariant Q(i) implies that no match exists at all.

We must now demonstrate that the algorithm never falsifies the invariant. It is easy to show that it is established at the beginning with the values i = j = 0. Let us first investigate the effect of the two statements incrementing i and j by 1. They apparently neither represent a shift of the pattern to the right, nor do they falsify Q(i-j), since the difference remains unchanged. But could they falsify P(i-j, j), the second factor of the invariant? We notice that at this point the negation of the inner while clause holds, i.e. either $j < 0$ or $s_i = p_j$. The latter extends the partial match and establishes P(i-j, j+1). In the former case, we postulate that P(i-j, j+1) hold as well. Hence, incrementing both i and j by 1 cannot falsify the invariant either. The only other assignment left in the algorithm is j := D. We shall simply postuate that the value D always be such that replacing j by D will maintain the invariant.

In order to find an appropriate expression for D, we must first understand the effect of the assignment. Provided that $D < j$, it represents a *shift of the pattern to the right* by j-D positions. Naturally, we wish this shift to be as large as possible, i.e., D to be as small as possible. This is illustrated by Fig. 1.10

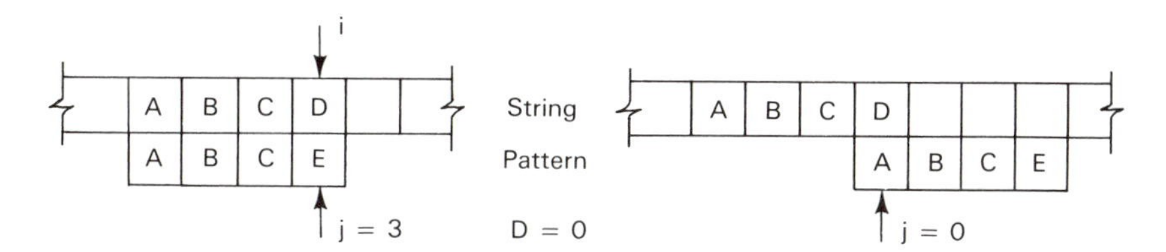

Fig. 1.10 Assignment j := D shifts pattern by j-D positions.

Evidently the condition P(i-D, D) & Q(i-D) must hold before assigning D to j, if the invariant P(i-j, j) & Q(i-j) is to hold thereafter. This precondition is therefore our guideline for finding an appropriate expression for D. The key observation is that thanks to P(i-j, j) we know that

$$s_{i-j} \cdots s_{i-1} = p_0 \cdots p_{j-1}$$

(we had just scanned the first j characters of the pattern and found them to match). Therefore the condition P(i-D, D) with $D \leq j$, i.e.,

$$p_0 \cdots p_{D-1} = s_{i-D} \cdots s_{i-1}$$

translates into

$$p_0 \cdots p_{D-1} = p_{j-D} \cdots p_{j-1} \qquad (1.51)$$

and (for the purpose of establishing the invariance of Q(i-D)) the predicate ~P(i-k, M) for k = 1 ... j-D translates into

$$p_0 \dots p_{k-1} \neq p_{j-k} \dots p_{j-1} \quad \text{for all } k = 1 \dots j\text{-}D \tag{1.52}$$

The essential result is that the value D apparently is determined by the pattern alone and does not depend on the text string. The conditions (1.51) and (1.52) tell us that in order to find D we must, for every j, search for the smallest D, and hence for the longest sequence of pattern characters just preceding position j, which matches an equal number of characters at the beginning of the pattern. We shall denote D for a given j by d_j. Since these values depend on the pattern only, the auxiliary table d may be computed before starting the actual search; this computation amounts to a *precompilation* of the pattern. This effort is evidently only worthwhile if the text is considerably longer than the pattern (M << N). If multiple occurrences of the same pattern are to be found, the same values of d can be reused. The following examples illustrate the function of d.

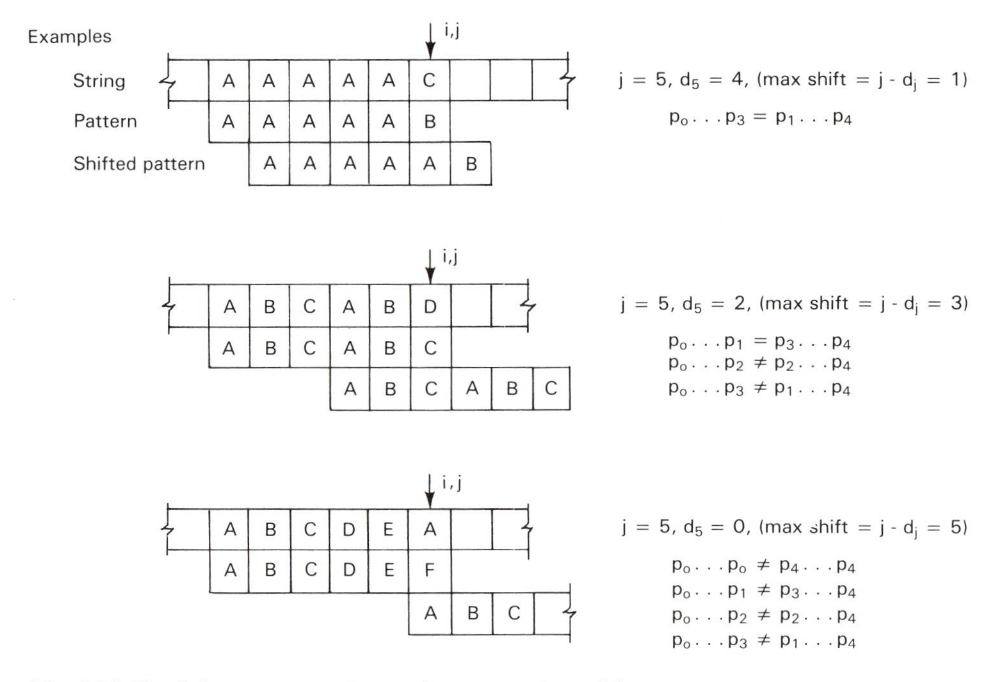

Fig. 1.11 Partial pattern matches and computation of d_j

The last example in Fig. 1.11 suggests that we can do even slightly better; had the the character p_j been an A instead of an F, we would know that the corresponding string character could not possibly be an A, because $s_i \neq p_j$ terminated the loop. Hence a shift of 5 could not lead to a later match either, and we might as well increase the shift amount to 6 (see Fig 1.11, upper part). Taking this into consideration, we redefine the computation of d_j

as the search for the longest matching sequence

$$p_0 \cdots p_{d_j-1} = p_{j-d_j} \cdots p_{j-1}$$

with the additional constraint of $p_{d_j} \neq p_j$. If no match exists at all, we let $d_j = -1$, indicating that the entire pattern be shifted beyond its current position (see Fig. 1.12, lower part).

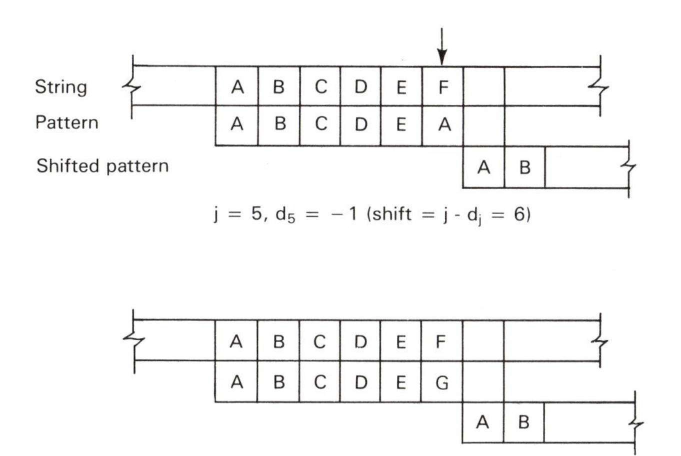

Fig. 1.12 Shifting pattern past position of last character.

Evidently, the computation of d_j presents us with the first application of string search, and we may as well use the fast KMP version itself. It is shown in Program 1.2, which consists of the following parts: First, the string s is read; then follows a repetition starting with the reading of a pattern, followed by the precompilation of the pattern, and finally by the search itself.

```
MODULE KMP;
  FROM InOut IMPORT
    OpenInput, CloseInput, Read, Write, WriteLn, Done;
  CONST Mmax = 100; Nmax = 10000; ESC = 33C;
  VAR i, j, k, k0, M, N: INTEGER;
      ch: CHAR;
      p: ARRAY [0 .. Mmax-1] OF CHAR;  (*pattern*)
      s: ARRAY [0 .. Nmax-1] OF CHAR;  (*string*)
      d: ARRAY [0 .. Mmax-1] OF INTEGER;
BEGIN OpenInput("TEXT"); N := 0; Read(ch);
  WHILE Done DO
    Write(ch); s[N] := ch; N := N+1; Read(ch)
  END ;
```

```
CloseInput;
LOOP WriteLn; Write(">"); M := 0; Read(ch);
  WHILE ch > " " DO
    Write(ch); p[M] := ch; M := M+1; Read(ch)
  END ;
  WriteLn;
  IF ch = ESC THEN EXIT END ;
  j := 0; k := -1; d[0] := -1;
  WHILE j < M-1 DO
    WHILE (k >= 0) & (p[j] # p[k]) DO k := d[k] END ;
    j := j+1; k := k+1;
    IF p[j] = p[k] THEN d[j] := d[k] ELSE d[j] := k END
  END ;
  i := 0; j := 0; k := 0;
  WHILE (j < M) & (i < N) DO
    WHILE k <= i DO Write(s[k]); k := k+1 END ;
    WHILE (j >= 0) & (s[i] # p[j]) DO j := d[j] END ;
    i := i+1; j := j+1
  END ;
  IF j = M THEN Write("!") (*found*) END
 END
END KMP.
```

Program 1.2 Knuth-Morris-Pratt String Search

Analysis of KMP search. The exact analysis of the performance of KMP-search is, like the algorithm itself, very intricate. In [1-8] its inventors prove that the number of character comparisons is in the order of M+N, which suggests a substantial improvement over M*N for the straight search. They also point out the welcome property that the scanning pointer i never backs up, whereas in straight string search the scan always begins at the first pattern character after a mismatch, and therefore may involve characters that had actually been scanned already. This may cause awkward problems when the string is read from secondary storage where backing up is costly. Even when the input is buffered, the pattern may be such that the backing up extends beyond the buffer contents.

1.12.6 Boyer-Moore String Search

The clever scheme of the KMP-search yields genuine benefits only if a mismatch was preceded by a partial match of some length. Only in this case is the pattern shift increased to more than 1. Unfortunately, this is the exception rather than the rule; matches occur much more seldom than mismatches. Therefore the gain in using the KMP strategy is marginal in most cases of normal text searching. The method to be discussed here does indeed not only improve performance in the worst case, but also in the average case. It was invented by R.S. Boyer and J.S. Moore around 1975, and we shall call it *BM search.* We shall here present a simplified version of BM-search before proceeding to the one given by Boyer and Moore..

BM-search is based on the unconventional idea to start comparing characters at the *end* of the pattern rather than at the beginning. Like in the case of KMP-search, the pattern is precompiled into a table d before the actual search starts. Let, for every character x in the character set, d_x be the distance of the rightmost occurrence of x in the pattern from its end. Now assume that a mismatch between string and pattern was discovered. Then the pattern can immediately be shifted to the right by $d_{p_{M-1}}$ positions, an amount that is quite likely to be greater than 1. If p_{M-1} does not occur in the pattern at all, the shift is even greater, namely equal to the entire pattern's length. The following example illustrates this process.

```
Hoola-Hoola girls like Hooligans.
Hooligan
     Hooligan
       Hooligan
               Hooligan
                       Hooligan
```

Since individual character comparisons now proceed from right to left, the following, slightly modified versions of of the predicates P and Q are more convenient.

$$P(i,j) = Ak: j \leq k < M : s_{i-j+k} = p_k \qquad (1.53)$$
$$Q(i) = Ak: 0 \leq k < i : \sim P(i, 0)$$

These predicates are used in the following formulation of the BM-algorithm to denote the invariant conditions.

```
i := M; j := M;
WHILE (j > 0) & (i < N) DO                                   (1.54)
  (* Q(i-M) *) j := M; k := i;
  WHILE (j > 0) & (s[k-1] = p[j-1]) DO
    (* P(k-j, j) & (k-j = i-M) *)
    k := k-1; j := j-1
  END ;
  i := i + d[s[i-1]]
END
```

The indices satisfy $0 \leq j \leq M$ and $0 \leq i,k \leq N$. Therefore, termination with $j = 0$, together with P(k-j, j), implies P(k,0), i.e., a match at position k. Termination with $j > 0$ demands that $i = N$; hence Q(i-M) implies Q(N-M), signalling that no match exists. Of course we still have to convince ourselves that Q(i-M) and P(k-j, j) are indeed invariants of the two repetitions. They are trivially satisfied when repetition starts, since Q(0) and P(x,M) are always true.

Let us first consider the effect of the two statements decrementing k and j. Q(i-M) is not affected, and, since $s_{k-1} = p_{j-1}$ had been established, P(k-j, j-1) holds as precondition, guaranteeing P(k-j, j) as postcondition. If the inner loop terminates with $j > 0$, the fact that $s_{k-1} \neq p_{j-1}$ implies ~P(k-j, 0), since

$$\sim P(i, 0) \;=\; \mathbf{E}k: 0 \leq k < M : s_{i+k} \neq p_k$$

Moreover, because k-j = M-i, Q(i-M) & ~P(k-j, 0) = Q(i+1-M), establishing a non-match at position i-M+1.

Next we must show that the statement $i := i + d_{s_{i-1}}$ never falsifies the invariant. This is the case, provided that before the assignment $Q(i+d_{s_{i-1}}-M)$ is guaranteed. Since we know that Q(i+1-M) holds, it suffices to establish ~P(i+h-M) for h = 2, 3, ... , $d_{s_{i-1}}$. We now recall that d_x is defined as the distance of the rightmost occurrence of x in the pattern from the end. This is formally expressed as

$$\mathrm{A}k: M-d_x \leq k < M-1 : p_k \neq x$$

Substituting s_i for x, we obtain

$$\mathrm{A}h: M-d_{s_{i-1}} \leq h < M-1 : s_{i-1} \neq p_h$$

$$\mathrm{A}h: 1 < h \leq d_{s_{i-1}} : s_{i-1} \neq p_{h-M}$$

$$\mathrm{A}h: 1 < h \leq d_{s_{i-1}} : \sim P(i+h-M)$$

The following program includes the presented, simplified Boyer-Moore strategy in a setting similar to that of the preceding KMP-search program. Note as a detail that a repeat statement is used in the inner loop, incrementing k and j before comparing s and p. This eliminates the -1 terms in the index expressions.

```
MODULE BM;
 FROM InOut IMPORT
  OpenInput, CloseInput, Read, Write, WriteLn, Done;

 CONST Mmax = 100; Nmax = 10000;

 VAR i, j, k, i0, M, N: INTEGER;
    ch: CHAR;
    p: ARRAY [0 .. Mmax-1] OF CHAR;  (*pattern*)
    s: ARRAY [0 .. Nmax-1] OF CHAR;  (*string*)
    d: ARRAY [0C .. 177C] OF INTEGER;

BEGIN OpenInput("TEXT"); N := 0; Read(ch);
 WHILE Done DO
  Write(ch); s[N] := ch; N := N+1; Read(ch)
 END ;
 CloseInput;
 LOOP WriteLn; Write(">"); M := 0; Read(ch);
  WHILE ch > " " DO
   Write(ch); p[M] := ch; M := M+1; Read(ch)
  END ;
  WriteLn;
  IF ch = 33C THEN EXIT END ;
```

```
    FOR ch := 0C TO 177C DO d[ch] := M END ;
    FOR j := 0 TO M-2 DO d[p[j]] := M-j-1 END ;

    i := M; i0 := 0;
    REPEAT
      WHILE i0 < i DO Write(s[i0]); i0 := i0+1 END ;
      j := M; k := i;
      REPEAT k := k-1; j := j-1
      UNTIL (j < 0) OR (p[j] # s[i]);
      i := i + d[s[i-1]]
    UNTIL (j < 0) OR (i >= N);
    IF j < 0 THEN Write("!") END
  END
END BM.
```

Program 1.3 Simplified Boyer-Moore String Search

Analysis of Boyer-Moore Search. The original publication of this algorithm [1-9] contains a detailed analysis of its performance. The remarkable property is that in all except especially construed cases it requires substantially less than N comparisons. In the luckiest case, where the last character of the pattern always hits an unequal character of the text, the number of comparisons is N/M.

The authors provide several ideas on possible further improvements. One is to combine the strategy explained above, which provides greater shifting steps when a *mismatch* is present, with the Knuth-Morris-Pratt strategy, which allows larger shifts after detection of a (partial) *match.* This method requires two precomputed tables; d1 is the table used above, and d2 is the table corresponding to the one of the KMP-algorithm. The step taken is then the larger of the two, both indicating that no smaller step could possibly lead to a match. We refrain from further elaborating the subject, because the additional complexity of the table generation and the search itself does not seem to yield any appreciable efficiency gain. In fact, the additional overhead is larger, and casts some uncertainty whether the sophisticated extension is an improvement or a deterioration.

EXERCISES

1.1. Assume that the cardinalities of the standard types INTEGER, REAL, and CHAR are denoted by c_I, c_R, and c_{CH}. What are the cardinalities of the following data types defined as exemples in this chapter: sex, BOOLEAN, weekday, letter, digit, officer, row, alfa, complex, date, person, coordinate, tapestatus?

1.2. How would you represent variables of the types listed in Exercise 1.1:

(a) In the store of your computer?
(b) In FORTRAN?
(c) In your favorite programming language?

1.3. Which are the instruction sequences (on your computer) for the following:

(a) Fetch and store operations for elements of packed records and arrays?
(b) Set operations, including the test for membership?

1.4. Can the correct use of variant records be checked at run time? Can it even be verified at compile time?

1.5. What are the reasons for defining certain sets of data as sequences instead of arrays?

1.6. Given is a railway timetable listing the daily services on several lines of a railway system. Find a representation of these data in terms of arrays, records, or sequences, which is suitable for lookup of arrival and departure times, given a certain station and desired direction of the train.

1.7. Given a text T in the form of a sequence and lists of a small number of words in the form of two arrays A and B. Assume that words are short arrays of characters of a small and fixed maximum length. Write a program that transforms the text T into a text S by replacing each occurrence of a word A_i by its corresponding word B_i.

1.8. Compare the following three versions of the binary search with (1.43). Which of the three programs are correct? Determine the relevant invariants. Which versions are more efficient? We assume the following variables, and the constant $N > 0$:

```
VAR i, j, k, x: INTEGER;
    a: ARRAY[1 .. N] OF INTEGER;
```

Program A:

```
i := 1; j := N;
REPEAT k := (i+j) DIV 2;
   IF a[k] < x THEN i := k ELSE j := k END
UNTIL (a[k] = x) OR (i ≥ j)
```

Program B:

```
i := 1; j := N;
REPEAT k := (i+j) DIV 2;
```

```
      IF x <= a[k] THEN j := k-1 END ;
      IF a[k] <= x THEN i := k+1 END
   UNTIL i > j
```

Program C:

```
   i := 1; j := N;
   REPEAT k := (i+j) DIV 2;
      IF x < a[k] THEN j := k ELSE i := k+1 END
   UNTIL i >= j
```

Hint: All programs must terminate with $a_k = x$, if such an element exists, or $a_k \neq x$, if there exists no element with value x.

1.9. A company organizes a poll to determine the success of its products. Its products are records and tapes of hits, and the most popular hits are to be broadcast in a hit parade. The polled population is to be divided into four categories according to sex and age (say, less or equal to 20, and older than 20). Every person is asked to name five hits. Hits are identified by the numbers 1 to N (say, N = 30). The results of the poll are to be appropriately encoded as a sequence of characters. Hint: use procedures *Read* and *ReadInt* to read the values of the poll.

```
TYPE hit = [0 .. N-1];
     sex = (male, female);
     reponse =
       RECORD  name, firstname: alfa;
               s: sex;
               age: INTEGER;
               choice: ARRAY [0 .. 4] OF hit
       END ;
VAR poll: Sequence
```

This file is the input to a program which computes the following results:

1. A list of hits in the order of their popularity. Each entry consists of the hit number and the number of times it was mentioned in the poll. Hits that were never mentioned are omitted from the list.

2. Four separate lists with the names and first names of all respondents who had mentioned in first place one of the three hits most popular in their category.

The five lists are to be preceded by suitable titles.

REFERENCES

1-1. O-.J. Dahl, E.W. Dijkstra, and C.A.R. Hoare. *Structured Programming.* (New York: Academic Press, 1972), pp. 155-65.

1-2. C.A.R. Hoare. Notes on data structuring; in *Structured Programming.* Dahl, Dijkstra, and Hoare, pp. 83-174.

1-3. K. Jensen and N. Wirth. *Pascal User Manual and Report.* (Berlin: Springer-Verlag, 1974).

1-4. N. Wirth. Program development by stepwise refinement. *Comm. ACM, 14,* No. 4 (1971), 221-27.

1-5. ------*Programming in Modula-2.* (Berlin, New York: Springer-Verlag, 1982).

1-6. ------, On the composition of well-structured programs.*Computing Surveys, 6,* No. 4, (1974) 247-59.

1-7. C.A.R. Hoare. The Monitor: A operating systems structuring concept. *Comm. ACM, 17,* 10 (Oct. 1974), 549-557.

1-8. D.E.Knuth, J.H. Morris, and V.R. Pratt. Fast pattern matching in strings. *SIAM J. Comput., 6,* 2, (June 1977), 323-349.

1-9. R.S. Boyer and J.S. Moore. A fast string searching algorithm. *Comm. ACM, 20,* 10 (Oct. 1977), 762-772.

2 Sorting

2.1. INTRODUCTION

The primary purpose of this chapter is to provide an extensive set of examples illustrating the use of the data structures introduced in the preceding chapter and to show how the choice of structure for the underlying data profoundly influences the algorithms that perform a given task. Sorting is also a good example to show that such a task may be performed according to many different algorithms, each one having certain advantages and disadvantages that have to be weighed against each other in the light of the particular application.

Sorting is generally understood to be the process of rearranging a given set of objects in a specific *order.* The purpose of sorting is to facilitate the later search for members of the sorted set. As such it is an almost universally performed, fundamental activity. Objects are sorted in telephone books, in income tax files, in tables of contents, in libraries, in dictionaries, in warehouses, and almost everywhere that stored objects have to be searched and retrieved. Even small children are taught to put their things "in order," and they are confronted with some sort of sorting long before they learn anything about arithmetic.

Hence, sorting is a relevant and essential activity, particularly in data processing. What else would be easier to sort than data! Nevertheless, our primary interest in sorting is devoted to the even more fundamental techniques used in the construction of algorithms. There are not many techniques that do not occur somewhere in connection with sorting algorithms. In particular, sorting is an ideal subject to demonstrate a great diversity of algorithms, all having the same purpose, many of them being optimal in some sense, and most of them having advantages over others. It is therefore an ideal subject to demonstrate the necessity of performance analysis of algorithms. The example of sorting is moreover well suited for showing how a very significant gain in performance may be obtained by the development of sophisticated algorithms when obvious methods are readily available.

The dependence of the choice of an algorithm on the structure of the data to be processed -- an ubiquitous phenomenon -- is so profound in the case of sorting that sorting methods are generally classified into two categories, namely, *sorting of arrays* and *sorting of (sequential) files*. The two classes are often called *internal* and *external* sorting because arrays are stored in the fast, high-speed, random-access "internal" store of computers and files are conveniently located on the slower, but more spacious "external" stores based on mechanically moving devices (disks and tapes). The importance of this distinction is obvious from the example of sorting numbered cards. Structuring the cards as an array corresponds to laying them out in front of the sorter so that each card is visible and individually accessible (see Fig. 2.1).

Fig. 2.1 Array sorting.

Structuring the cards as a file, however, implies that from each pile only the card on the top is visible (see Fig. 2.2). Such a restriction will evidently have serious consequences on the sorting method to be used, but it is unavoidable if the number of cards to be laid out is larger than the available table.

Before proceeding, we introduce some terminology and notation to be used throughout this chapter. If we are given items

$$a_1, a_2, \dots, a_n$$

sorting consists of permuting these items into an array

$$a_{k_1}, a_{k_2}, \dots, a_{k_n}$$

such that, given an ordering function f,

$$f(a_{k_1}) \le f(a_{k_2}) \le \dots \le f(a_{k_n}). \qquad (2.1.)$$

Ordinarily, the ordering function is not evaluated according to a specified rule of computation but is stored as an explicit component (field) of each item. Its value is called the *key* of the item. As a consequence, the record structure is particularly well suited to

Fig. 2.2 File sorting.

represent items and might for example be declared as follows:

```
TYPE item =  RECORD key: INTEGER;
               (*other components declared here*)          (2.2)
             END
```

The other components represent relevant data about the items in the collection; the key merely assumes the purpose of identifying the items. As far as our sorting algorithms are concerned, however, the key is the only relevant component, and there is no need to define any particular remaining components. In the following discussions, we shall therefore discard any associated information and assume that the type item be defined as INTEGER. The choice of INTEGER as the key type is somewhat arbitrary. Evidently, any type on which a total ordering relation is defined could be used just as well.

A sorting method is called *stable* if the relative order if items with equal keys remains unchanged by the sorting process. Stability of sorting is often desirable, if items are already ordered (sorted) according to some secondary keys, i.e., properties not reflected by the (primary) key itself.

This chapter is not to be regarded as a comprehensive survey in sorting techniques. Rather, some selected, specific methods are exemplified in greater detail. For a thorough treatment of sorting, the interested reader is referred to the excellent and comprehensive compendium by D. E. Knuth [2-7] (see also Lorin [2-10]).

2.2. SORTING ARRAYS

The predominant requirement that has to be made for sorting methods on arrays is an economical use of the available store. This implies that the permutation of items which brings the items into order has to be performed *in situ,* and that methods which transport items from an array *a* to a result array *b* are intrinsically of minor interest. Having thus restricted our choice of methods among the many possible solutions by the criterion of economy of storage, we proceed to a first classification according to their efficiency, i.e., their economy of time. A good measure of efficiency is obtained by counting the numbers C of needed key comparisons and M of moves (transpositions) of items. These numbers are functions of the number *n* of items to be sorted. Whereas good sorting algorithms require in the order of n * log n comparisons, we first discuss several simple and obvious sorting techniques, called *straight methods,* all of which require in the order n^2 comparisons of keys. There are three good reasons for presenting straight methods before proceeding to the faster algorithms.

1. Straight methods are particularly well suited for elucidating the characteristics of the major sorting principles.
2. Their programs are easy to understand and are short. Remember that programs occupy storage as well!
3. Although sophisticated methods require fewer operations, these operations are usually more complex in their details; consequently, straight methods are faster for sufficiently small n, although they must not be used for large n.

Sorting methods that sort items *in situ* can be classified into three principal categories according to their underlying method:

1. Sorting by insertion.
2. Sorting by selection.
3. Sorting by exchange.

These three pinciples will now be examined and compared. The programs operate on the variable *a* whose components are to be sorted in situ and refer to the data types *item* (2.2) and *index,* defined as

```
TYPE index = INTEGER;
VAR a: ARRAY[1 .. n] OF item                                    (2.3)
```

2.2.1. Sorting by Straight Insertion

This method is widely used by card players. The items (cards) are conceptually divided into a destination sequence $a_1 \dots a_{i-1}$ and a source sequence $a_i \dots a_n$. In each step, starting with i = 2 and incrementing i by unity, the i th element of the source sequence is picked and transferred into the destination sequence by inserting it at the appropriate place.

Initial Keys	44	55	12	42	94	18	06	67
i=2	44	55	12	42	94	18	06	67
i=3	12	44	55	42	94	18	06	67
i=4	12	42	44	55	94	18	06	67
i=5	12	42	44	55	94	18	06	67
i=6	12	18	42	44	55	94	06	67

i=7	06	12	18	42	44	55	94	67
i=8	06	12	18	42	44	55	67	94

Table 2.1 A Sample Process of Straight Insertion Sorting.

The process of sorting by insertion is shown in an example of eight numbers chosen at random (see Table 2.1). The algorithm of straight insertion is

```
FOR i := 2 TO n DO
  x := a[i];
  insert x at the appropriate place in a1 ... ai
END
```

In the process of actually finding the appropriate place, it is convenient to alternate between comparisons and moves, i.e., to let x sift down by comparing x with the next item a_j, and either inserting x or moving a_j to the right and proceeding to the left. We note that there are two distinct conditions that may cause the termination of the sifting down process:

1. An item a_j is found with a key less than the key of x.
2. The left end of the destination sequence is reached.

This typical case of a repetition with two termination conditions brings the well-known sentinel technique to our attention. It is easily applied to this case by posting a sentinel a_0 with the value of x. (Note that this requires the extension of the index range in the declaration of *a* to 0 .. n.) The completed algorithm is formulated in Program 2.1.

```
PROCEDURE StraightInsertion;
  VAR i, j: index; x: item;
BEGIN
  FOR i := 2 TO n DO
    x := a[i]; a[0] := x; j := i;
    WHILE x < a[j-1] DO a[j] := a[j-1]; j := j-1 END ;
    a[j] := x
  END
END StraightInsertion
```

Program 2.1 Sorting by Straight Insertion

Analysis of straight insertion. The number C_i of key comparisons in the i-th sift is at most i-1, at least 1, and -- assuming that all permutations of the n keys are equally probable -- i/2 in the average. The number M_i of moves (assignments of items) is $C_i + 2$ (including the sentinel). Therefore, the total numbers of comparisons and moves are

$$C_{min} = n-1 \qquad M_{min} = 3*(n-1) \qquad (2.4)$$
$$C_{ave} = (n^2 + n - 2)/4 \qquad M_{ave} = (n^2 + 9n - 10)/4$$
$$C_{max} = (n^2 + n - 4)/4 \qquad M_{max} = (n^2 + 3n - 4)/2$$

The minimal numbers occur if the items are initially in order; the worst case occurs if the items are initially in reverse order. In this sense, sorting by insertion exhibits a truly natural behavior. It is plain that the given algorithm also describes a stable sorting process: it leaves the order of items with equal keys unchanged.

The algorithm of straight insertion is easily improved by noting that the destination sequence $a_i \dots a_{i-1}$, in which the new item has to be inserted, is already ordered. Therefore, a faster method of determining the insertion point can be used. The obvious choice is a binary search that samples the destination sequence in the middle and continues bisecting until the insertion point is found. The modified sorting algorithm is called *binary insertion*, and is shown as Program 2.2.

```
PROCEDURE BinaryInsertion;
  VAR i, j, m, L, R: index; x: item;
BEGIN
  FOR i := 2 TO n DO
    x := a[i]; L := 1; R := i;
    WHILE L < R DO
      m := (L+R) DIV 2;
      IF a[m] <= x THEN L := m+1 ELSE R := m END
    END ;
    FOR j := i TO R+1 BY -1 DO a[j] := a[j-1] END ;
    a[R] := x
  END
END BinaryInsertion
```

Program 2.2. Sorting by Binary Insertion.

Analysis of binary insertion. The insertion position is found if L = R. Thus, the search interval must in the end be of length 1; and this involves halving the interval of length i log i times. Thus,

$$C = \mathbf{S}\, i: 1 \leq i \leq n: \lceil \log i \rceil$$

We approximate this sum by the integral

$$\text{Integral } (1{:}n) \log x \, dx = n*(\log n - c) + c \qquad (2.5)$$

where $c = \log e = 1/\ln 2 = 1.44269\dots$. The number of comparisons is essentially independent of the initial order of the items. However, because of the truncating character of the division involved in bisecting the search interval, the true number of comparisons needed with i items may be up to 1 higher than expected. The nature of this bias is such that insertion positions at the low end are on the average located slightly faster than those at the high end, thereby favoring those cases in which the items are originally highly out of order. In fact, the minimum number of comparisons is needed if the items are initially in reverse order and the maximum if they are already in order. Hence, this is a case of unnatural behavior of a sorting algorithm.

$$C \doteq n*(\log n - \log e \pm 0.5)$$

Unfortunately, the improvement obtained by using a binary search method applies only to the number of comparisons but not to the number of necessary moves. In fact, since moving items, i.e., keys and associated information, is in general considerably more time-consuming than comparing two keys, the improvement is by no means drastic: the important term M is still of the order n^2. And, in fact, sorting the already sorted array takes more time than does straight insertion with sequential search.

This example demonstrates that an "obvious improvement" often has much less drastic consequences than one is first inclined to estimate and that in some cases (that do occur) the "improvement" may actually turn out to be a deterioration. After all, sorting by insertion does not appear to be a very suitable method for digital computers: insertion of an item with the subsequent shifting of an entire row of items by a single position is uneconomical. One should expect better results from a method in which moves of items are only performed upon single items and over longer distances. This idea leads to sorting by selection.

2.2.2 Sorting by Straight Selection

This method is based on the following principle:

1. Select the item with the least key.
2. Exchange it with the first item a_1.
3. Then repeat these operations with the remaining n-1 items, then with n-2 items, until only one item -- the largest -- is left.

This method is shown on the same eight keys as in Table 2.1.

Initial keys	44	55	12	42	94	18	06	67
	06	55	12	42	94	18	44	67
	06	12	55	42	94	18	44	67
	06	12	18	42	94	55	44	67
	06	12	18	42	94	55	44	67
	06	12	18	42	44	55	94	67
	06	12	18	42	44	55	94	67
	06	12	18	42	44	55	67	94

Table 2.2 A Sample Process of Straight Selection Sorting.

The algorithm is formulated as follows:

FOR i := 1 TO n-1 DO
  assign the index of the least item of $a_i \dots a_n$ *to k;*
  exchange a_i *with* a_k
END

This method, called *straight selection,* is in some sense the opposite of straight insertion: Straight insertion considers in each step only the *one* next item of the source sequence and *all* items of the destination array to find the insertion point; straight selection considers *all* items of the source array to find the *one* with the least key and to be deposited as the one next item of the destination sequence. The entire program of straight selection is given in

Program 2.3.

```
PROCEDURE StraightSelection;
  VAR i, j, k: index; x: item;
BEGIN
  FOR i := 1 TO n-1 DO
    k := i; x := a[i];
    FOR j := i+1 TO n DO
      IF a[j] < x THEN k := j; x := a[k] END
    END ;
    a[k] := a[i]; a[i] := x
  END
END StraightSelection
```

Analysis of straight selection. Evidently, the number C of key comparisons is independent of the initial order of keys. In this sense, this method may be said to behave less naturally than straight insertion. We obtain

$$C = (n^2 - n)/2$$

The number M of moves is at least

$$M_{min} = 3*(n-1) \tag{2.6}$$

in the case of initially ordered keys and at most

$$M_{max} = n^2/4 + 3*(n-1)$$

if initially the keys are in reverse order. In order to determine M_{avg} we make the following deliberations: The algorithm scans the array, comparing each element with the minimal value so far detected and, if smaller than that minimum, performs an assignment. The probability that the second element is less than the first, is 1/2; this is also the probability for a new assignment to the minimum. The chance for the third element to be less than the first two is 1/3, and the chance of the fourth to be the smallest is 1/4, and so on. Therefore the total expected number of moves is H_n-1, where H_n is the n th harmonic number

$$H_n = 1 + 1/2 + 1/3 + ... + 1/n \tag{2.7}$$

H_n can be expressed as

$$H_n = \ln n + g + 1/2n - 1/12n^2 + ... \tag{2.8}$$

where $g = 0.577216...$ is Euler's constant. For sufficiently large n, we may ignore the fractional terms and therefore approximate the average number of assignments in the i th pass as

$$F_i = \ln i + g + 1$$

The average number of moves M_{avg} in a selection sort is then the sum of F_i with i ranging from 1 to n.

$$M_{avg} = n*(g+1) + (\mathbf{S}\, i: 1 \le i \le n: \ln i)$$

By further approximating the sum of discrete terms by the integral

$$\text{Integral } (1:n) \ln x \, dx = x * (\ln x - 1) = n * \ln(n) - n + 1$$

we obtain an approximate value

$$M_{avg} \doteq n * (\ln(n) + g) \tag{2.9}$$

We may conclude that in general the algorithm of straight selection is to be preferred over straight insertion, although in the cases in which keys are initially sorted or almost sorted, straight insertion is still somewhat faster.

2.2.3 Sorting by Straight Exchange

The classification of a sorting method is seldom entirely clear-cut. Both previously discussed methods can also be viewed as exchange sorts. In this section, however, we present a method in which the exchange of two items is the dominant characteristic of the process. The subsequent algorithm of straight exchanging is based on the principle of comparing and exchanging pairs of adjacent items until all items are sorted.

As in the previous methods of straight selection, we make repeated passes over the array, each time sifting the least item of the remaining set to the left end of the array. If, for a change, we view the array to be in a vertical instead of a horizontal position, and -- with the help of some imagination -- the items as bubbles in a water tank with weights according to their keys, then each pass over the array results in the ascension of a bubble to its appropriate level of weight (see Table 2.3). This method is widely known as the *Bubblesort.* Its simplest form is shown in Program 2.4.

i=1	2	3	4	5	6	7	8
44	06	06	06	06	06	06	06
55	44	12	12	12	12	12	12
12	55	44	18	18	18	18	18
42	12	55	44	42	42	42	42
94	42	18	55	44	44	44	44
18	94	42	42	55	55	55	55
06	18	94	67	67	67	67	67
67	67	67	94	94	94	94	94

Table 2.3 A Sample of Bubblesorting.

```
PROCEDURE BubbleSort;
  VAR i, j: index; x: item;
BEGIN
  FOR i := 2 TO n DO
    FOR j := n TO i BY -1 DO
      IF a[j-1] > a[j] THEN
```

```
        x := a[j-1]; a[j-1] := a[j]; a[j] := x
      END
    END
  END
END BubbleSort
```

Program 2.4 Bubblesort

This algorithm easily lends itself to some improvements. The example in Table 2.3 shows that the last three passes have no effect on the order of the items because the items are already sorted. An obvious technique for improving this algorithm is to remember whether or not any exchange had taken place during a pass. A last pass without further exchange operations is therefore necessary to determine that the algorithm may be terminated. However, this improvement may itself be improved by remembering not merely the fact that an exchange took place, but rather the position (index) of the last exchange. For example, it is plain that all pairs of adjacent items below this index k are in the desired order. Subsequent scans may therefore be terminated at this index instead of having to proceed to the predetermined lower limit i. The careful programmer notices, however, a peculiar asymmetry: A single misplaced bubble in the heavy end of an otherwise sorted array will sift into order in a single pass, but a misplaced item in the light end will sink towards its correct position only one step in each pass. For example, the array

12 18 42 44 55 67 94 06

is sorted by the improved Bubblesort in a single pass, but the array

94 06 12 18 42 44 55 67

requires seven passes for sorting. This unnatural asymmetry suggests a third improvement: alternating the direction of consecutive passes. We appropriately call the resulting algorithm *Shakersort.* Its behavior is illustrated in Table 2.4 by applying it to the same eight keys that were used in Table 2.3.

```
PROCEDURE ShakerSort;
  VAR j, k, L, R: index;  x: item;
BEGIN L := 2; R := n; k := n;
  REPEAT
    FOR j := R TO L BY -1 DO
      IF a[j-1] > a[j] THEN
        x := a[j-1]; a[j-1] := a[j]; a[j] := x; k := j
      END
    END ;
    L := k+1;
    FOR j := L TO R BY +1 DO
      IF a[j-1] > a[j] THEN
        x := a[j-1]; a[j-1] := a[j]; a[j] := x; k := j
      END
    END ;
    R := k-1
```

```
    UNTIL L > R
  END ShakerSort
```

Program 2.5 Shakersort

L=	2	3	3	4	4
R=	8	8	7	7	4
dir=	↑	↓	↑	↓	↑
	44	06	06	06	06
	55	44	44	12	12
	12	55	12	44	18
	42	12	42	18	42
	94	42	55	42	44
	18	94	18	55	55
	06	18	67	67	67
	67	67	94	94	94

Table 2.4 An Example of Shakersort.

Analysis of Bubblesort and Shakersort. The number of comparisons in the straight exchange algorithm is

$$C = (n^2 - n)/2 \tag{2.10}$$

and the minimum, average, and maximum numbers of moves (assignments of items) are

$$M_{min} = 0, \qquad M_{avg} = 3*(n^2 - n)/2, \qquad M_{max} = 3*(n^2 - n)/4 \tag{2.11}$$

The analysis of the improved methods, particularly that of Shakersort, is intricate. The least number of comparisons is $C_{min} = n-1$. For the improved Bubblesort, Knuth arrives at an average number of passes proportional to $n - k_1 n^{½}$, and an average number of comparisons proportional to $½(n^2 - n(k_2 + \ln n))$. But we note that all improvements mentioned above do in no way affect the number of exchanges; they only reduce the number of redundant double checks. Unfortunately, an exchange of two items is generally a more costly operation than a comparison of keys; our clever improvements therefore have a much less profound effect than one would intuitively expect.

This analysis shows that the exchange sort and its minor improvements are inferior to both the insertion and the selection sorts; and in fact, the Bubblesort has hardly anything to recommend it except its catchy name. The Shakersort algorithm is used with advantage in those cases in which it is known that the items are already almost in order -- a rare case in practice.

It can be shown that the average distance that each of the n items has to travel during a sort is n/3 places. This figure provides a clue in the search for improved, i.e. more effective sorting methods. All straight sorting methods essentially move each item by one position in each elementary step. Therefore, they are bound to require in the order n^2 such steps. Any improvement must be based on the principle of moving items over greater distances in single leaps.

Subsequently, three improved methods will be discussed, namely, one for each basic sorting method: insertion, selection, and exchange.

2.3. ADVANCED SORTING METHODS

2.3.1 Insertion Sort by Diminishing Increment

A refinement of the straight insertion sort was proposed by D. L. Shell in 1959. The method is explained and demonstrated on our standard example of eight items (see Table 2.5). First, all items that are four positions apart are grouped and sorted separately. This process is called a 4-sort. In this example of eight items, each group contains exactly two items. After this first pass, the items are regrouped into groups with items two positions apart and then sorted anew. This process is called a 2-sort. Finally, in a third pass, all items are sorted in an ordinary sort or 1-sort.

One may at first wonder if the necessity of several sorting passes, each of which involves all items, does not introduce more work than it saves. However, each sorting step over a chain either involves relatively few items or the items are already quite well ordered and comparatively few rearrangements are required.

It is obvious that the method results in an ordered array, and it is fairly obvious that each pass profits from previous passes (since each i-sort combines two groups sorted in the preceding 2i-sort). It is also obvious that any sequence of increments is acceptable, as long as the last one is unity, because in the worst case the last pass does all the work. It is, however, much less obvious that the method of diminishing increments yields even better results with increments other than powers of 2.

	44	55	12	42	94	18	06	67
4-sort yields								
	44	18	06	42	94	55	12	67
2-sort yield								
	06	18	12	42	44	55	94	67
1-sort yields								
	06	12	18	42	44	55	67	94

Table 2.5 An Insertion Sort with Diminishing Increments.

The program is therefore developed without relying on a specific sequence of increments. The t increments are denoted by

$$h_1, h_2, \dots, h_t$$

with the conditions

$$h_t = 1, \ \ h_{i+1} < h_i \tag{2.12}$$

Each h-sort is programmed as a straight insertion sort using the sentinel technique to provide a simple termination condition for the search of the insertion place. It is plain that each sort needs to post its own sentinel and that the program to determine its position should be made as simple as possible. The array a therefore has to be extended not only by a single component a_0, but by h_1 components, such that it is now declared as

a: ARRAY [$-h_1$ .. n] OF item

The algorithm is described by the procedure called *Shellsort* [2.11] in Program 2.6 for t = 4.

```
PROCEDURE ShellSort;
  CONST t = 4;
  VAR i, j, k, s: index;
    x: item;  m: 1 .. t;
    h: ARRAY [1 .. t] OF INTEGER;
BEGIN h[1] := 9; h[2] := 5; h[3] := 3; h[4] := 1;
  FOR m := 1 TO t DO
    k := h[m]; s := -k;  (*sentinel position*)
    FOR i := k+1 TO n DO
      x := a[i]; j := i-k;
      IF s = 0 THEN s := -k END ;
      s := s+1; a[s] := x;  (*post sentinel*)
      WHILE x < a[j] DO a[j+k] := a[j]; j := j-k END ;
      a[j+k] := x
    END
  END
END ShellSort
```

Program 2.6. Shellsort.

Analysis of Shellsort. The analysis of this algorithm poses some very difficult mathematical problems, many of which have not yet been solved. In particular, it is not known which choice of increments yields the best results. One surprising fact, however, is that they should not be multiples of each other. This will avoid the phenomenon evident from the example given above in which each sorting pass combines two chains that before had no interaction whatsoever. It is indeed desirable that interaction between various chains takes place as often as possible, and the following theorem holds: If a k-sorted sequence is i-sorted, then it remains k-sorted. Knuth [2.8] indicates evidence that a reasonable choice of increments is the sequence (written in reverse order)

1, 4, 13, 40, 121, ...

where $h_{k-1} = 3h_k+1$, $h_t = 1$, and $t = \lfloor \log_3 n \rfloor - 1$. He also recommends the sequence

1, 3, 7, 15, 31, ...

where $h_{k-1} = 2h_k+1$, $h_t = 1$, and $t = \lfloor \log_2 n \rfloor - 1$. For the latter choice, mathematical analysis yields an effort proportional to $n^{1.2}$ required for sorting n items with the Shellsort algorithm. Although this is a significant improvement over n^2, we will not expound further on this method, since even better algorithms are known.

2.3.2 Tree Sort

The method of sorting by straight selection is based on the repeated selection of the least key among n items, then among the remaining n-1 items, etc. Clearly, finding the least key among n items requires n-1 comparisons, finding it among n-1 items needs n-2 comparisons, etc., and the sum of the first n-1 integers is $\frac{1}{2}(n^2-n)$. So how can this selection sort possibly be improved? It can be improved only by retaining from each scan more information than just the identification of the single least item. For instance, with n/2 comparisons it is possible to determine the smaller key of each pair of items, with another n/4 comparisons the smaller of each pair of such smaller keys can be selected, and so on. With only n-1 comparisons, we can construct a selection tree as shown in Fig. 2.3. and identify the root as the desired least key [2.2].

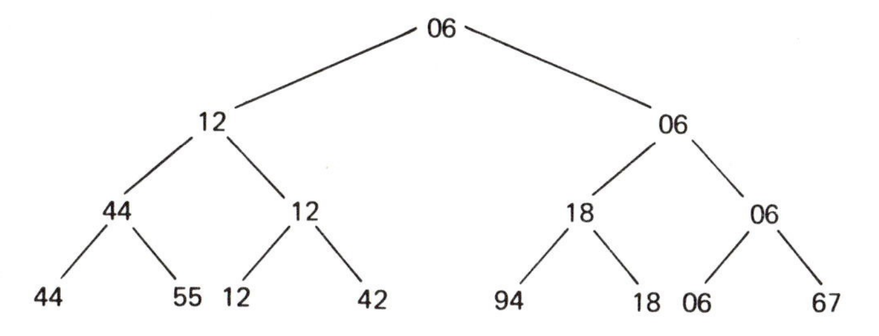

Fig. 2.3 Repeated selection between two keys.

The second step now consists of descending down along the path marked by the least key and eliminating it by successively replacing it by either an empty hole at the bottom, or by the item at the alternative branch at intermediate nodes (see Figs. 2.4 and 2.5). Again, the item emerging at the root of the tree has the (now second) smallest key and can be eliminated. After n such selection steps, the tree is empty (i.e., full of holes), and the sorting process is terminated. It should be noted that each of the n selection steps requires only log n comparisons. Therefore, the total selection process requires only on the order of n*log n elementary operations in addition to the n steps required by the construction of the tree. This is a very significant improvement over the straight methods requiring n^2 steps, and even over Shellsort that requires $n^{1.2}$ steps. Naturally, the task of bookkeeping has become more elaborate, and therefore the complexity of individual steps is greater in the tree sort method; after all, in order to retain the increased amount of information gained from the initial pass, some sort of tree structure has to be created. Our next task is to find methods of organizing this information efficiently.

Of course, it would seem particularly desirable to eliminate the need for the holes that in the end populate the entire tree and are the source of many unnecessary comparisons. Moreover, a way should be found to represent the tree of n items in n units of storage, instead of in 2n - 1 units as shown above. These goals are indeed achieved by a method called *Heapsort* by its inventor J. Williams [2-14]; it is plain that this method represents a

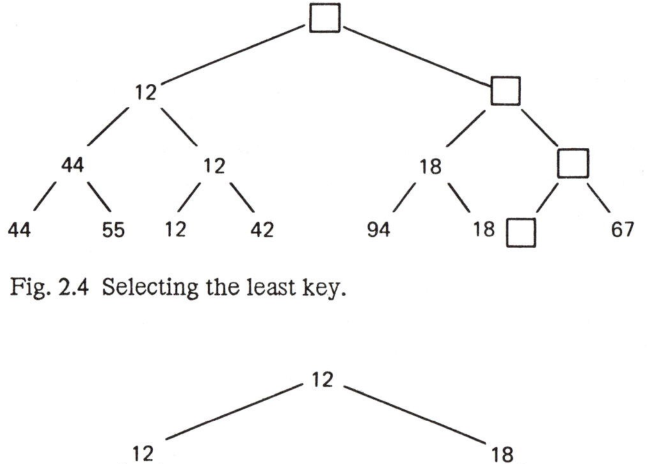

Fig. 2.4 Selecting the least key.

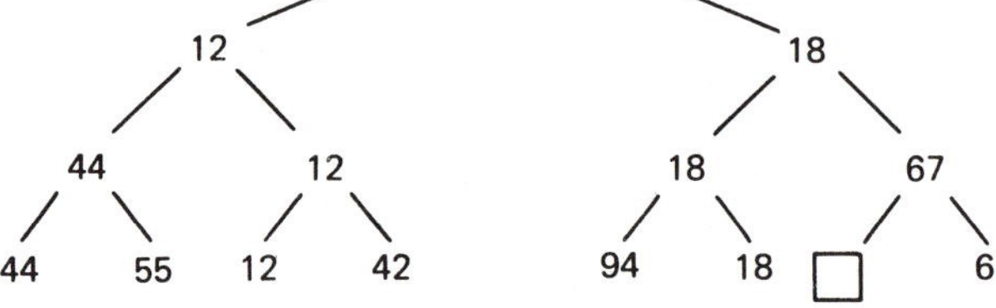

Fig. 2.5 Refilling the holes.

drastic improvement over more conventional tree sorting approaches. A *heap* is defined as a sequence of keys $h_L, h_{L+1}, \dots, h_R$ such that

$$h_i \leq h_{2i} \quad \text{and} \quad h_i \leq h_{2i+1} \quad \text{for } i = L \dots R/2. \tag{2.13}$$

If a binary tree is represented as an array as shown in Fig. 2.6, then it follows that the sort trees in Figs. 2.7 and 2.8 are heaps, and in particular that the element h_1 of a heap is its least element:

$$h_1 = \min(h_1, h_2, \dots, h_n)$$

Let us now assume that a heap with elements $h_{L+1} \dots h_R$ is given for some values L and R, and that a new element x has to be added to form the extended heap $h_L \dots h_R$. Take, for example, the initial heap $h_1 \dots h_7$ shown in Fig. 2.7 and extend the heap to the left by an element $h_1 = 44$. A new heap is obtained by first putting x on top of the tree structure and then by letting it sift down along the path of the smaller comparands, which at the same time move up. In the given example the value 44 is first exchanged with 06, then with 12, and thus forming the tree shown in Fig. 2.8. We now formulate this sifting algorithm as follows: i, j are the pair of indices denoting the items to be exchanged during each sift step. The reader is urged to convince himself that the proposed method of sifting actually preserves the conditions (2.13) that define a heap.

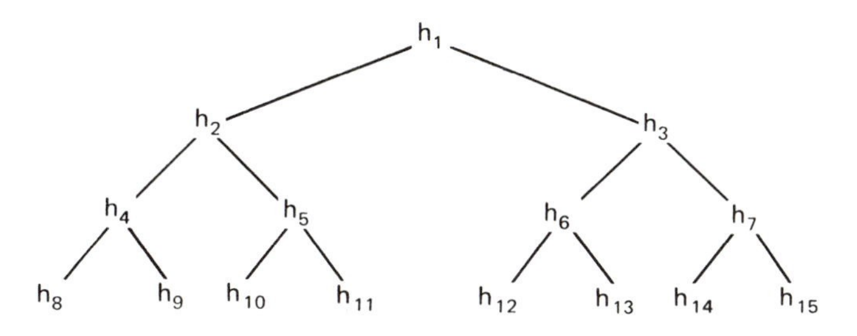

Fig. 2.6 Array h viewed as a binary tree.

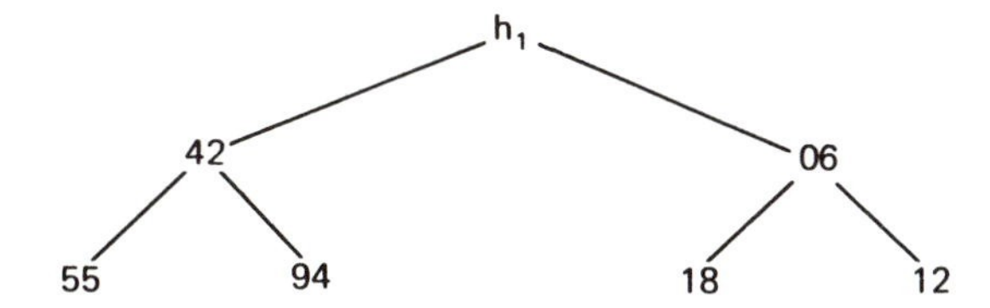

Fig. 2.7 Heap with seven elements.

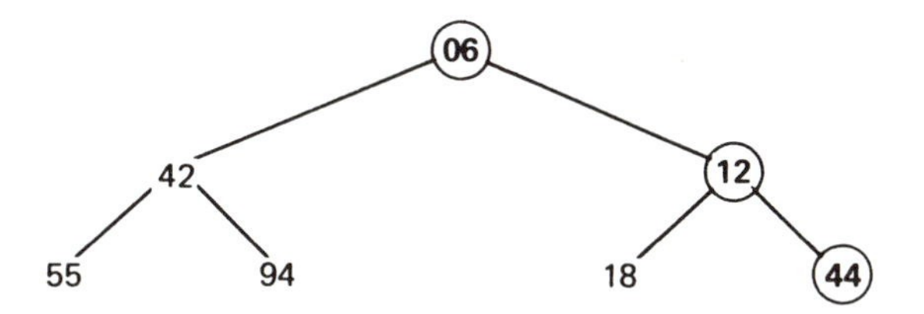

Fig. 2.8 Key 44 sifting through the heap.

A neat way to construct a heap *in situ* was suggested by R. W. Floyd. It uses the sifting procedure shown in Program 2.7. Given is an array $h_1 \dots h_n$; clearly, the elements $h_m \dots h_n$ (with m = (n DIV 2)+1) form a heap already, since no two indices i, j are such that j = 2i (or j = 2i+1). These elements form what may be considered as the bottom row of the associated binary tree (see Fig. 2.6) among which no ordering relationship is required. The heap is now extended to the left, whereby in each step a new element is included and properly positioned by a sift. This process is illustrated in Table 2.6 and yields the heap shown in Fig. 2.6.

```
PROCEDURE sift(L, R: index);
  VAR i, j: index;  x: item;
BEGIN i := L; j := 2*L; x := a[L];
  IF (j < R) & (a[j+1] < a[j]) THEN j := j+1 END ;
  WHILE (j <= R) & (a[j] < x) DO
    a[i] := a[j]; i := j; j := 2*j;
```

```
    IF (j < R) & (a[j+1] < a[j]) THEN j := j+1 END
  END
END sift
```

Program 2.7 Sift.

<table>
<tr><td>44</td><td>55</td><td>12</td><td>42 |</td><td>94</td><td>18</td><td>06</td><td>67</td></tr>
<tr><td>44</td><td>55</td><td>12 |</td><td>42</td><td>94</td><td>18</td><td>06</td><td>67</td></tr>
<tr><td>44</td><td>55 |</td><td>06</td><td>42</td><td>94</td><td>18</td><td>12</td><td>67</td></tr>
<tr><td>44 |</td><td>42</td><td>06</td><td>55</td><td>94</td><td>18</td><td>12</td><td>67</td></tr>
<tr><td>06</td><td>42</td><td>12</td><td>55</td><td>94</td><td>18</td><td>44</td><td>67</td></tr>
</table>

Table 2.6 Constructing a Heap.

Consequently, the process of generating a heap of n elements $h_1 \dots h_n$ *in situ* is described as follows:

```
L := (n DIV 2) + 1;
WHILE L > 1 DO  L := L-1; sift(L, n) END
```

In order to obtain not only a partial, but a full ordering among the elements, n sift steps have to follow, whereby after each step the next (least) item may be picked off the top of the heap. Once more, the question arises about where to store the emerging top elements and whether or not an *in situ* sort would be possible. Of course there is such a solution: In each step take the last component (say x) off the heap, store the top element of the heap in the now free location of x, and let x sift down into its proper position. The necessary n-1 steps are illustrated on the heap of Table 2.7. The process is described with the aid of the procedure *sift* (Program 2.7) as follows:

```
R := n;
WHILE R > 1 DO
  x := a[1]; a[1] := a[R]; a[R] := x;
  R := R-1; sift(1, R)
END
```

<table>
<tr><td>06</td><td>42</td><td>12</td><td>55</td><td>94</td><td>18</td><td>44</td><td>67</td></tr>
<tr><td>12</td><td>42</td><td>18</td><td>55</td><td>94</td><td>67</td><td>44 |</td><td>06</td></tr>
<tr><td>18</td><td>42</td><td>44</td><td>55</td><td>94</td><td>67 |</td><td>12</td><td>06</td></tr>
<tr><td>42</td><td>55</td><td>44</td><td>67</td><td>94 |</td><td>18</td><td>12</td><td>06</td></tr>
<tr><td>44</td><td>55</td><td>94</td><td>67 |</td><td>42</td><td>18</td><td>12</td><td>06</td></tr>
<tr><td>55</td><td>67</td><td>94 |</td><td>44</td><td>42</td><td>18</td><td>12</td><td>06</td></tr>
<tr><td>67</td><td>94 |</td><td>55</td><td>44</td><td>42</td><td>18</td><td>12</td><td>06</td></tr>
<tr><td>94 |</td><td>67</td><td>55</td><td>44</td><td>42</td><td>18</td><td>12</td><td>06</td></tr>
</table>

Table 2.7 Example of a Heapsort Process.

The example of Table 2.7 shows that the resulting order is actually inverted. This, however, can easily be remedied by changing the direction of the ordering relations in the *sift* procedure. This results in the procedure Heapsort shown in Program 2.8.

```
PROCEDURE HeapSort;
  VAR L, R: index;  x: item;

  PROCEDURE sift(L, R: index);
    VAR i, j: index;  x: item;
  BEGIN i := L; j := 2*L; x := a[L];
    IF (j < R) & (a[j] < a[j+1]) THEN j := j+1 END ;
    WHILE (j <= R) & (x < a[j]) DO
      a[i] := a[j]; i := j; j := 2*j;
      IF (j < R) & (a[j] < a[j+1]) THEN j := j+1 END
    END ; a[i] := x
  END sift;

BEGIN L := (n DIV 2) + 1; R := n;
  WHILE L > 1 DO L := L-1; sift(L, R) END ;
  WHILE R > 1 DO
    x := a[1]; a[1] := a[R]; a[R] := x;
    R := R-1; sift(L, R)
  END
END HeapSort
```

Program 2.8 Heapsort.

Analysis of Heapsort. At first sight it is not evident that this method of sorting provides good results. After all, the large items are first sifted to the left before finally being deposited at the far right. Indeed, the procedure is not recommended for small numbers of items, such as shown in the example. However, for large n, Heapsort is very efficient, and the larger n is, the better it becomes -- even compared to Shellsort.

In the worst case, there are n/2 sift steps necessary, sifting items through log(n/2), log(n/2 -1), ... , log(n-1) positions, where the logarithm (to the base 2) is truncated to the next lower integer. Subsequently, the sorting phase takes n-1 sifts, with at most log(n-1), log(n-2), ... , 1 moves. In addition, there are n-1 moves for stashing the sifted item away at the right. This argument shows that Heapsort takes of the order of n*log n steps even in the worst possible case. This excellent worst-case performance is one of the strongest qualities of Heapsort.

It is not at all clear in which case the worst (or the best) performance can be expected. But generally Heapsort seems to like initial sequences in which the items are more or less sorted in the inverse order, and therefore it displays an unnatural behavior. The heap creation phase requires zero moves if the inverse order is present. The average number of moves is approximately n/2 * log(n), and the deviations from this value are relatively small.

2.3.3 Partition Sort

After having discussed two advanced sorting methods based on the principles of insertion and selection, we introduce a third improved method based on the principle of exchange. In view of the fact that Bubblesort was on the average the least effective of the three straight sorting algorithms, a relatively significant improvement factor should be expected. Still, it comes as a surprise that the improvement based on exchanges to be discussed subsequently yields the best sorting method on arrays known so far. Its performance is so spectacular that its inventor, C.A.R. Hoare, called it *Quicksort* [2.5 and 2.6].

Quicksort is based on the fact that exchanges should preferably be performed over large distances in order to be most effective. Assume that n items are given in reverse order of their keys. It is possible to sort them by performing only n/2 exchanges, first taking the leftmost and the rightmost and gradually progressing inward from both sides. Naturally, this is possible only if we know that their order is exactly inverse. But something might still be learned from this example.

Let us try the following algorithm: Pick any item at random (and call it x); scan the array from the left until an item $a_i > x$ is found and then scan from the right until an item $a_j < x$ is found. Now exchange the two items and continue this *scan and swap* process until the two scans meet somewhere in the middle of the array. The result is that the array is now partitioned into a left part with keys less than (or equal to) x, and a right part with keys greater than (or equal to) x. This partitioning process is now formulated in the form of a procedure in Program 2.9. Note that the relations > and < have been replaced by $\geq$ and $\leq$, whose negations in the while clause are < and >. With this change x acts as a sentinel for both scans.

```
PROCEDURE partition;
  VAR w, x: item;
BEGIN i := 1; j := n;
  select an item x at random;
  REPEAT
    WHILE a[i] < x DO i := i+1 END ;
    WHILE x < a[j] DO j := j-1 END ;
    IF i <= j THEN
      w := a[i]; a[i] := a[j]; a[j] := w; i := i+1; j := j-1
    END
  UNTIL i > j
END partition
```

Program 2.9 Partition.

As an example, if the middle key 42 is selected as comparand x, then the array of keys

44 55 12 42 94 06 18 67

requires the two exchanges 18↔44 and 6↔55 to yield the partitioned array

18 06 12 42 94 55 44 67

and the final index values i = 5 and j = 3. Keys $a_1 \ldots a_{i-1}$ are less or equal to key x = 42, and keys $a_{j+1} \ldots a_n$ are greater or equal to key x. Consequently, there are three partitions, namely

$$Ak : 1 \leq k < i : a_k \leq x \qquad (2.14)$$
$$Ak : j < k \leq n : x \leq a_k$$

This algorithm is very straightforward and efficient because the essential comparands i, j, and x can be kept in fast registers throughout the scan. However, it can also be cumbersome, as witnessed by the case with n identical keys, which result in n/2 exchanges. These unnecessary exchanges might easily be eliminated by changing the scanning statements to

```
WHILE a[i] <= x DO i := i+1 END ;
WHILE x <= a[j] DO j := j-1 END
```

In this case, however, the choice element x, which is present as a member of the array, no longer acts as a sentinel for the two scans. The array with all identical keys would cause the scans to go beyond the bounds of the array unless more complicated termination conditions were used. The simplicity of the conditions used in Program 2.9 is well worth the extra exchanges that occur relatively rarely in the average random case. A slight saving, however, may be achieved by changing the clause controlling the exchange step to $i < j$ instead of $i \leq j$. But this change must not be extended over the two statements

```
i := i+1; j := j-1
```

which therefore require a separate conditional clause. Confidence in the correctness of the partition algorithm can be gained by verifying that the relations (2.14) are invariants of the repeat statement. Initially, with $i = 1$ and $j = n$, they are trivially true, and upon exit with $i > j$, they imply the desired result.

We now recall that our goal is not only to find partitions of the original array of items, but also to sort it. However, it is only a small step from partitioning to sorting: after partitioning the array, apply the same process to both partitions, then to the partitions of the partitions, and so on, until every partition consists of a single item only. This recipe is described by Program 2.10.

```
PROCEDURE QuickSort;

   PROCEDURE sort(L, R: index);
     VAR i, j: index; w, x: item;
   BEGIN i := L; j := R;
     x := a[(L+R) DIV 2];
     REPEAT
       WHILE a[i] < x DO i := i+1 END ;
       WHILE x < a[j] DO j := j-1 END ;
       IF i <= j THEN
         w := a[i]; a[i] := a[j]; a[j] := w; i := i+1; j := j-1
       END
     UNTIL i > j;
     IF L < j THEN sort(L, j) END ;
     IF i < R THEN sort(i, R) END
```

```
    END sort;

  BEGIN sort(1, n)
  END QuickSort
```

Program 2.10. Quicksort.

Procedure *sort* activates itself recursively. Such use of recursion in algorithms is a very powerful tool and will be discussed further in Chap. 3. In some programming languages of older provenience, recursion is disallowed for certain technical reasons. We will now show how this same algorithm can be expressed as a non-recursive procedure. Obviously, the solution is to express recursion as an iteration, whereby a certain amount of additional bookkeeping operations become necessary.

The key to an iterative solution lies in maintaining a list of partitioning requests that have yet to be performed. After each step, two partitioning tasks arise. Only one of them can be attacked directly by the subsequent iteration; the other one is stacked away on that list. It is, of course, essential that the list of requests is obeyed in a specific sequence, namely, in reverse sequence. This implies that the first request listed is the last one to be obeyed, and vice versa; the list is treated as a pulsating stack. In the following nonrecursive version of Quicksort, each request is represented simply by a left and a right index specifying the bounds of the partition to be further partitioned. Thus, we introduce an array variable called *stack* and an index *s* designating its most recent entry (see Program 2.11). The appropriate choice of the stack size M will be discussed during the analysis of Quicksort.

```
PROCEDURE NonRecursiveQuickSort;
  CONST M = 12;
  VAR i, j, L, R: index;  x, w: item;
    s: [0 .. M];
    stack: ARRAY [1 .. M] OF RECORD L, R: index END ;
BEGIN s := 1; stack[1].L := 1; stack[s].R := n;
  REPEAT (*take top request from stack*)
    L := stack[s].L; R := stack[s].R; s := s-1;
    REPEAT (*partition a[L] ... a[R]*)
      i := L; j := R; x := a[(L+R) DIV 2];
      REPEAT
        WHILE a[i] < x DO i := i+1 END ;
        WHILE x < a[j] DO j := j-1 END ;
        IF i <= j THEN
          w := a[i]; a[i] := a[j]; a[j] := w; i := i+1; j := j-1
        END
      UNTIL i > j;
      IF i < R THEN  (*stack request to sort right partition*)
        s := s+1; stack[s].L := i; stack[s].R := R
      END ;
      R := j  (*now L and R delimit the left partition*)
    UNTIL L >= R
  UNTIL s = 0
```

END NonRecursiveQuickSort

Program 2.11 Non-recursive Version of Quicksort.

Analysis of Quicksort. In order to analyze the performance of Quicksort, we need to investigate the behavior of the partitioning process first. After having selected a bound x, it sweeps the entire array. Hence, exactly n comparisons are performed. The number of exchanges can be determind by the following probabilistic argument.

With a fixed bound x, the expected number of exchange operations is equal to the number of elements in the left part of the partition, namely n-1, multiplied by the probability that such an element reached its place by an exchange. An exchange had taken place if the element had previously been part of the right partition; the probablity for this is (n-(x-1))/n. The expected number of exchanges is therefore the average of these expected values over all possible bounds x.

$$\begin{aligned} M &= [Sx: 1 \leq x \leq n : (x-1)*(n-(x-1))/n]/n \\ &= [Su: 0 \leq u \leq n-1 : u*(n-u)]/n^2 \\ &= n*(n-1)/2n - (2n^2 - 3n + 1)/6n \;=\; (n - 1/n)/6 \end{aligned} \qquad (2.15)$$

Assuming that we are very lucky and always happen to select the median as the bound, then each partitioning process splits the array in two halves, and the number of necessary passes to sort is log n. The resulting total number of comparisons is then n*log n, and the total number of exchanges is n * log(n)/6.

Of course, one cannot expect to hit the median all the time. In fact, the chance of doing so is only 1/n. Surprisingly, however, the average performance of Quicksort is inferior to the optimal case by a factor of only 2*ln(2), if the bound is chosen at random.

But Quicksort does have its pitfalls. First of all, it performs moderately well for small values of n, as do all advanced methods. Its advantage over the other advanced methods lies in the ease with which a straight sorting method can be incorporated to handle small partitions. This is particularly advantageous when considering the recursive version of the program.

Still, there remains the question of the worst case. How does Quicksort perform then? The answer is unfortunately disappointing and it unveils the one weakness of Quicksort. Consider, for instance, the unlucky case in which each time the largest value of a partition happens to be picked as comparand x. Then each step splits a segment of n items into a left partition with n-1 items and a right partition with a single element. The result is that n (instead of log n) splits become necessary, and that the worst-case performance is of the order n^2.

Apparently, the crucial step is the selection of the comparand x. In our example program it is chosen as the middle element. Note that one might almost as well select either the first or the last element. In these cases, the worst case is the initially sorted array; Quicksort then shows a definite dislike for the trivial job and a preference for disordered arrays. In choosing the middle element, the strange characteristic of Quicksort is less obvious because the initially sorted array becomes the optimal case. In fact, also the average performance is slightly better, if the middle element is selected. Hoare suggests that the choice of x be made at random, or by selecting it as the median of a small sample of, say, three keys [2.12 and 2.13]. Such a judicious choice hardly influences the average performance of Quicksort, but it improves the worst-case performance considerably. It becomes evident that sorting on the basis of Quicksort is somewhat like a gamble in which one should be aware of how much one may afford to lose if bad luck were to strike.

There is one important lesson to be learned from this experience; it concerns the programmer directly. What are the consequences of the worst case behavior mentioned above to the performance of Program 2.11? We have realized that each split results in a right partition of only a single element; the request to sort this partition is stacked for later execution. Consequently, the maximum number of requests, and therefore the total required stack size, is n. This is, of course, totally unacceptable. (Note that we fare no better -- in fact even worse -- with the recursive version because a system allowing recursive activation of procedures will have to store the values of local variables and parameters of all procedure activations automatically, and it will use an implicit stack for this purpose.)

The remedy lies in stacking the sort request for the *longer* partition and in continuing directly with the further partitioning of the smaller section. In this case, the size of the stack M can be limited to log n.

The change necessary to Program 2.11 is localized in the section setting up new requests. It now reads

```
IF j - L < R - i THEN
  IF i < R THEN (*stack request for sorting right partition*)
    s := s+1; stack[s].L := i; stack[s].R := R
  END ;
  R := j (*continue sorting left partition*)
ELSE                                                          (2.16)
  IF L < j THEN (*stack request for sorting left parition*)
    s := s+1; stack[s].L := L; srack[s].R := j
  END;
  L := i (*continue sorting right partition*)
END
```

2.3.4. Finding the Median

The *median* of n items is defined as that item which is less than (or equal to) half of the n items and which is larger than (or equal to) the other half of the n items. For example, the median of

16 12 99 95 18 87 10

is 18. The problem of finding the median is customarily connected with that of sorting, because the obvious method of determining the median is to sort the n items and then to pick the item in the middle. But partitioning by Program 2.9 yields a potentially much faster way of finding the median. The method to be displayed easily generalizes to the problem of finding the k th smallest of n items. Finding the median represents the special case $k = n/2$.

The algorithm invented by C.A.R. Hoare [2-4] functions as follows. First, the partitioning operation of Quicksort is applied with $L = 1$ and $R = n$ and with a_k selected as splitting value x. The resulting index values i and j are such that

$$\begin{array}{ll} 1.\ a_h < x & \text{for all } h < i \\ 2.\ a_h > x & \text{for all } h > j \\ 3.\ i > j & \end{array} \qquad (2.17)$$

There are three possible cases that may arise:

1. The splitting value x was too small; as a result, the limit between the two partitions is below the desired value k. The partitioning process has to be repeated upon the elements $a_i \dots a_R$ (see Fig. 2.9).

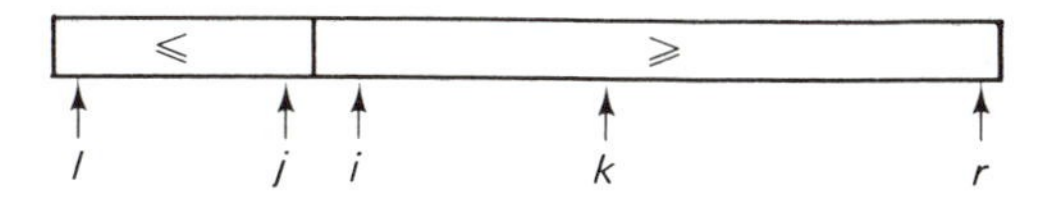

Fig. 2.9 Bound too small.

2. The chosen bound x was too large. The splitting operation has to be repeated on the partition $a_L \dots a_j$ (see Fig. 2.10).

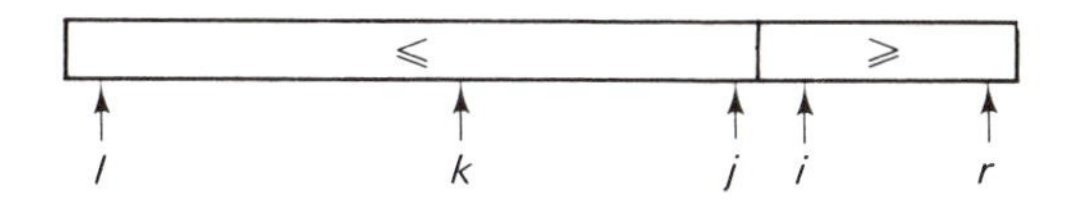

Fig. 2.10 Bound too large.

3. $j < k < i$: the element a_k splits the array into two partitions in the specified proportions and therefore is the desired quantile (see Fig. 2.11).

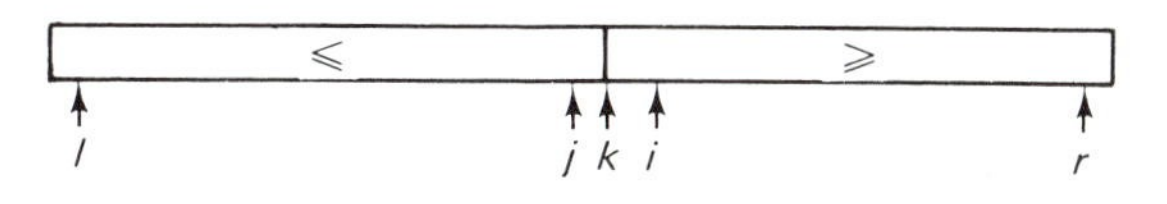

Fig. 2.11 Correct bound.

The splitting process has to be repeated until case 3 arises. This iteration is expressed by the following piece of program:

```
L := 1; R := n;
WHILE L < R DO                                              (2.18)
  x := a[k]; partition(a[L] ... a[R]);
  IF j < k THEN L := i END ;
  IF k < i THEN R := j END
END
```

For a formal proof of the correctness of this algorithm, the reader is referred to the original article by Hoare. The entire program *Find* is readily derived from this.

```
PROCEDURE Find(k: INTEGER);
  VAR L, R, i, j: index; w, x: item;
BEGIN L := 1; R := n;
  WHILE L < R DO
    x := a[k]; i := L; j := R;
    REPEAT
      WHILE a[i] < x DO i := i+1 END ;
      WHILE x < a[j] DO j := j-1 END ;
      IF i <= j THEN
        w := a[i]; a[i] := a[j]; a[j] := w; i := i+1; j := j-1
      END
    UNTIL i > j;
    IF j < k THEN L := i END ;
    IF k < i THEN R := j END
  END
END Find
```

Program 2.12. Find the k th largest element.

If we assume that on the average each split halves the size of the partition in which the desired quantile lies, then the number of necessary comparisons is

$$n + n/2 + n/4 + \ldots + 1 \doteq 2n \tag{2.19}$$

i.e., it is of order n. This explains the power of the program *Find* for finding medians and similar quantiles, and it explains its superiority over the straightforward method of sorting the entire set of candidates before selecting the k th (where the best is of order $n * \log(n)$). In the worst case, however, each partitioning step reduces the size of the set of candidates only by 1, resulting in a required number of comparisons of order n^2. Again, there is hardly any advantage in using this algorithm, if the number of elements is small, say, fewer than 10.

2.3.5. A Comparison of Array Sorting Methods

To conclude this parade of sorting methods, we shall try to compare their effectiveness. If n denotes the number of items to be sorted, C and M shall again stand for the number of required key comparisons and item moves, respectively. Closed analytical formulas can be given for all three straight sorting methods. They are tabulated in Table 2.8. The column headings Min, Avg, Max, specify the respective minima, maxima, and values averaged over all n! permutations of n items.

		Min	Avg	Max
Straight insertion	C =	n-1	$(n^2 + n - 2)/4$	$(n^2 - n)/2 - 1$
	M =	2(n-1)	$(n^2 - 9n - 10)/4$	$(n^2 - 3n - 4)/2$
Straight selection	C =	$(n^2 - n)/2$	$(n^2 - n)/2$	$(n^2 - n)/2$
	M =	3(n-1)	n*(ln n + 0.57)	$n^2/4 + 3(n-1)$
Straight exchange	C =	$(n^2-n)/2$	$(n^2-n)/2$	$(n^2-n)/2$
	M =	0	$(n^2-n)*0.75$	$(n^2-n)*1.5$

Table 2.8. Comparison of Straight Sorting Methods.

No reasonably simple accurate formulas are available on the advanced methods. The essential facts are that the computational effort needed is $c*n^{1.2}$ in the case of Shellsort and is c*n*log n in the cases of Heapsort and Quicksort, where the c are appropriate coefficients.

These formulas merely provide a rough measure of performance as functions of n, and they allow the classification of sorting algorithms into primitive, straight methods (n^2) and advanced or "logarithmic" methods (n*log(n)). For practical purposes, however, it is helpful to have some experimental data available that shed light on the coefficients c which further distinguish the various methods. Moreover, the formulas do not take into account the computational effort expended on operations other than key comparisons and item moves, such as loop control, etc. Clearly, these factors depend to some degree on individual systems, but an example of experimentally obtained data is nevertheless informative. Table 2.9 shows the times (in seconds) consumed by the sorting methods previously discussed, as executed by the Modula-2 system on a Lilith personal computer. The three columns contain the times used to sort the already ordered array, a random permutation, and the inversely ordered array. The left figure in each column is for 256 items, the right one for 512 items. The data clearly separate the n^2 methods from the n*log(n) methods. The following points are noteworthy:

1. The improvement of binary insertion over straight insertion is marginal indeed, and even negative in the case of an already existing order.
2. Bubblesort is definitely the worst sorting method among all compared. Its improved version Shakersort is still worse than straight insertion and straight selection (except in the pathological case of sorting a sorted array).
3. Quicksort beats Heapsort by a factor of 2 to 3. It sorts the inversely ordered array with speed practically identical to the one that is already sorted.

	Ordered	Random	Inverse
	n = 256		
StraightInsertion	0.02	0.82	1.64

BinaryInsertion	0.12	0.70	1.30
StraightSelection	0.94	0.96	1.18
BubbleSort	1.26	2.04	2.80
ShakerSort	0.02	1.66	2.92
ShellSort	0.10	0.24	0.28
HeapSort	0.20	0.20	0.20
QuickSort	0.08	0.12	0.08
NonRecQuickSort	0.08	0.12	0.08
StraightMerge	0.18	0.18	0.18

n = 2048

StraightInsertion	0.22	50.74	103.80
BinaryInsertion	1.16	37.66	76.06
StraightSelection	58.18	58.34	73.46
BubbleSort	80.18	128.84	178.66
ShakerSort	0.16	104.44	187.36
ShellSort	0.80	7.08	12.34
HeapSort	2.32	2.22	2.12
QuickSort	0.72	1.22	0.76
NonRecQuickSort	0.72	1.32	0.80
StraightMerge	1.98	2.06	1.98

Tables 2.9. and 2.10. Execution times of sort programs

2.4. SORTING SEQUENCES

2.4.1. Straight Merging

Unfortunately, the sorting algorithms presented in the preceding chapter are inapplicable, if the amount of data to be sorted does not fit into a computer's main store, but if it is, for instance, represented on a peripheral and sequential storage device such as a tape or a disk. In this case we describe the data as a (sequential) file whose characteristic is that at each moment one and only one component is directly accessible. This is a severe restriction compared to the possibilities offered by the array structure, and therefore different sorting techniques have to be used. The most important one is sorting by *merging.* Merging (or collating) means combining two (or more) ordered sequences into a single, ordered sequence by repeated selection among the currently accessible components. Merging is a much simpler operation than sorting, and it is used as an auxiliary operation in the more complex process of sequential sorting. One way of sorting on the basis of merging, called *straight merging,* is the following:

1. Split the sequence *a* into two halves, called *b* and *c.*
2. Merge *b* and *c* by combining single items into ordered pairs.
3. Call the merged sequence *a,* and repeat steps 1 and 2, this time merging ordered pairs into ordered quadruples.
4. Repeat the previous steps, merging quadruples into octets, and continue doing this, each time doubling the lengths of the merged subsequences, until the entire sequence is ordered.

As an example, consider the sequence

    44   55   12   42   94   18   06   67

In step 1, the split results in the sequences

    44   55   12   42
    94   18   06   67

The merging of single components (which are ordered sequences of length 1), into ordered pairs yields

    44   94 ' 18   55 ' 06   12 ' 42   67

Splitting again in the middle and merging ordered pairs yields

    06   12   44   94 ' 18   42   55   67

A third split and merge operation finally produces the desired result

    06   12   18   42   44   55   67   94

Each operation that treats the entire set of data once is called a *phase,* and the smallest subprocess that by repetition constitutes the sort process is called a *pass* or a *stage.* In the above example the sort took three passes, each pass consisting of a splitting phase and a merging phase. In order to perform the sort, three tapes are needed; the process is therefore called a *three-tape merge.*

Actually, the splitting phases do not contribute to the sort since they do in no way permute the items; in a sense they are unproductive, although they constitute half of all copying operations. They can be eliminated altogether by combining the split and the merge phase. Instead of merging into a single sequence, the output of the merge process is immediately redistributed onto two tapes, which constitute the sources of the subsequent pass. In contrast to the previous *two-phase merge* sort, this method is called a *single-phase merge* or a *balanced merge.* It is evidently superior because only half as many copying operations are necessary; the price for this advantage is a fourth tape.

We shall develop a merge program in detail and initially let the data be represented as an array which, however, is scanned in strictly sequential fashion. A later version of merge sort will then be based on the sequence structure, allowing a comparison of the two programs and demonstrating the strong dependence of the form of a program on the underlying representation of its data.

A single array may easily be used in place of two sequences, if it is regarded as double-ended. Instead of merging from two source files, we may pick items off the two ends of the array. Thus, the general form of the combined merge-split phase can be illustrated as shown in Fig. 2.12. The destination of the merged items is switched after each ordered pair in the first pass, after each ordered quadruple in the second pass, etc., thus evenly filling the two destination sequences, represented by the two ends of a single array. After each pass, the two arrays interchange their roles, the source becomes the new destination, and vice versa.

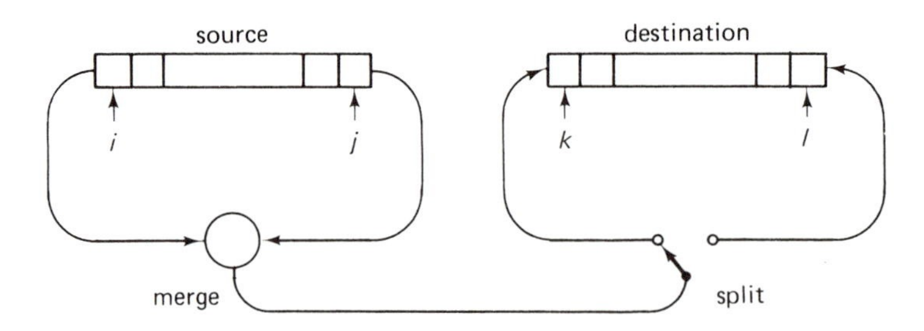

Fig. 2.12 Straight merge sort with two arrays.

A further simplification of the program can be achieved by joining the two conceptually distinct arrays into a single array of doubled size. Thus, the data will be represented by

a: ARRAY[1 .. 2*n] OF item (2.20)

and we let the indices i and j denote the two source items, whereas k and L designate the two destinations (see Fig. 2.12). The initial data are, of course, the items $a_1 \dots a_n$. Clearly, a

Boolean variable *up* is needed to denote the direction of the data flow; *up* shall mean that in the current pass components $a_1 \dots a_n$ will be moved up to the variables $a_{n+1} \dots a_{2n}$, whereas ~*up* will indicate that $a_{n+1} \dots a_{2n}$ will be transferred down into $a_1 \dots a_n$. The value of *up* strictly alternates between consecutive passes. And, finally, a variable p is introduced to denote the length of the subsequences to be merged. Its value is initially 1, and it is doubled before each successive pass. To simplify matters somewhat, we shall assume that n is always a power of 2. Thus, the first version of the straight merge program assumes the following form:

```
PROCEDURE MergeSort;                                              (2.21)
  VAR i, j, k, L: index;  up: BOOLEAN;  p: INTEGER;
BEGIN up := TRUE; p := 1;
  REPEAT initialize index variables;
    IF up THEN i := 1; j := n; k := n+1; L := 2*n
    ELSE k := 1; L := n; i := n+1; j := 2*n
    END ;
    merge p-tuples from i- and j-sources to k- and L-destinations;
    up := ~up; p := 2*p
  UNTIL p = n
END MergeSort
```

In the next development step we further refine the statements expressed in italics. Evidently, the merge pass involving n items is itself a sequence of merges of sequences, i.e. of p-tuples. Between every such partial merge the destination is switched from the lower to the upper end of the destination array, or vice versa, to guarantee equal distribution onto both destinations. If the destination of the merged items is the lower end of the destination array, then the destination index is k, and k is incremented after each move of an item. If they are to be moved to the upper end of the destination array, the destination index is L, and it is decremented after each move. In order to simplify the actual merge statement, we choose the destination to be designated by k at all times, switching the values of the variables k and L after each p-tuple merge, and denote the increment to be used at all times by h, where h is either 1 or -1. These design discussions lead to the following refinement:

```
h := 1; m := n;  (*m = no. of items to be merged*)               (2.22)
REPEAT q := p; r := p; m := m - 2*p;
  merge q items from i-source with r items from j-source.
  destination index is k. increment k by h;
  h := -h; k ↔ L
UNTIL m = 0
```

In the further refinement step the actual merge statement is to be formulated. Here we have to keep in mind that the tail of the one subsequence which is left non-empty after the merge has to be appended to the output sequence by simple copying operations.

```
WHILE (q # 0) & (r # 0) DO
  IF a[i] < a[j] THEN                                             (2.23)
    move an item from i-source to k-destination; advance i and k; q := q-1
```

```
      ELSE
        move an item from j-source to k-destination; advance j and k; r := r-1
      END
    END ;
    copy tail of i-sequence;copy tail of j-sequence
```

After this further refinement of the tail copying operations, the program is laid out in complete detail. Before writing it out in full, we wish to eliminate the restriction that n be a power of 2. Which parts of the algorithm are affected by this relaxation of constraints? We easily convince ourselves that the best way to cope with the more general situation is to adhere to the old method as long as possible. In this example this means that we continue merging p-tuples until the remainders of the source sequences are of length less than p. The one and only part that is influenced are the statements that determine the values of q and r, the lengths of the sequences to be merged. The following four statements replace the three statements

```
q := p;  r := p;  m := m -2*p
```

and, as the reader should convince himself, they represent an effective implementation of the strategy specified above; note that m denotes the total number of items in the two source sequences that remain to be merged:

```
IF m >= p THEN q := p ELSE q := m END ;
m := m-q;
IF m >= p THEN r := p ELSE r := m END ;
m := m-r
```

In addition, in order to guarantee termination of the program, the condition p=n, which controls the outer repetition, must be changed to $p \geq n$. After these modifications, we may now proceed to describe the entire algorithm in terms of a complete program (see Program 2.13).

```
PROCEDURE StraightMerge;
  VAR i, j, k, L, t: index;  (*index range is 1 .. 2*n*)
    h, m, p, q, r: INTEGER; up: BOOLEAN;
BEGIN up := TRUE; p := 1;
  REPEAT h := 1; m := n;
    IF up THEN i := 1; j := n; k := n+1; L := 2*n
    ELSE k := 1; L := n; i := n+1; j := 2*n
    END ;
    REPEAT (*merge a run from i- and j-sources to k-destination*)
      IF m >= p THEN q := p ELSE q := m END ;
      m := m-q;
      IF m >= p THEN r := p ELSE r := m END ;
      m := m-r;
      WHILE (q # 0) & (r # 0) DO
        IF a[i] < a[j] THEN
          a[k] := a[i]; k := k+h; i := i+1; q := q-1
        ELSE
          a[k] := a[j]; k := k+h; j := j-1; r := r-1
```

```
          END
        END ;
        WHILE r > 0 DO
          a[k] := a[j]; k := k+h; j := j-1; r := r-1
        END ;
        WHILE q > 0 DO
          a[k] := a[i]; k := k+h; i := i+1; q := q-1
        END ;
        h := -h; t := k; k := L; L := t
      UNTIL m = 0;
      up := ~up; p := 2*p
    UNTIL p >= n;
    IF ~up THEN
      FOR i := 1 TO n DO a[i] := a[i+n] END
    END
  END StraightMerge
```

Program 2.13 Straight Mergesort on Array.

Analysis of Mergesort. Since each pass doubles p, and since the sort is terminated as soon as $p \geq n$, it involves $\lceil \log n \rceil$ passes. Each pass, by definition, copies the entire set of n items exactly once. As a consequence, the total number of moves is exactly

$$M = n * \lceil \log n \rceil \tag{2.24}$$

The number C of key comparisons is even less than M since no comparisons are involved in the tail copying operations. However, since the mergesort technique is usually applied in connection with the use of peripheral storage devices, the computational effort involved in the move operations dominates the effort of comparisons often by several orders of magnitude. The detailed analysis of the number of comparisons is therefore of little practical interest.

The merge sort algorithm apparently compares well with even the advanced sorting techniques discussed in the previous chapter. However, the administrative overhead for the manipulation of indices is relatively high, and the decisive disadvantage is the need for storage of 2n items. This is the reason sorting by merging is rarely used on arrays, i.e., on data located in main store. Figures comparing the real time behavior of this Mergesort algorithm appear in the last line of Table 2.9. They compare favorably with Heapsort but unfavorably with Quicksort.

2.4.2. Natural Merging

In straight merging no advantage is gained when the data are initially already partially sorted. The length of all merged subsequences in the k th pass is less than or equal to 2^k, independent of whether longer subsequences are already ordered and could as well be merged. In fact, any two ordered subsequences of lengths m and n might be merged directly into a single sequence of m+n items. A mergesort that at any time merges the two longest possible subsequences is called a *natural merge sort.*

An ordered subsequence is often called a *string.* However, since the word string is even more frequently used to describe sequences of characters, we will follow Knuth in our terminology and use the word *run* instead of string when referring to ordered subsequences. We call a subsequence $a_i \dots a_j$ such that

$$(a_{i-1} > a_i) \mathbin{\&} (Ak : i \leq k < j : a_k \leq a_{k+1}) \mathbin{\&} (a_j > a_{j+1}) \tag{2.25}$$

a *maximal run* or, for short, a *run*. A natural merge sort, therefore, merges (maximal) runs instead of sequences of fixed, predetermined length. Runs have the property that if two sequences of n runs are merged, a single sequence of exactly n runs emerges. Therefore, the total number of runs is halved in each pass, and the number of required moves of items is in the worst case $n*\lceil \log n \rceil$, but in the average case it is even less. The expected number of comparisons, however, is much larger because in addition to the comparisons necessary for the selection of items, further comparisons are needed between consecutive items of each file in order to determine the end of each run.

Our next programming exercise develops a natural merge algorithm in the same stepwise fashion that was used to explain the straight merging algorithm. It employs the sequence structure instead of the array, and it represents an unbalanced, two-phase, three-tape merge sort. We assume that the variable c represents the initial sequence of items. (Naturally, in actual data processing application, the initial data are first copied from the original source to c for reasons of safety.) a and b are two auxiliary sequence variables. Each pass consists of a distribution phase that distributes runs equally from c to a and b, and a merge phase that merges runs from a and b to c. This process is illustrated in Fig. 2.13.

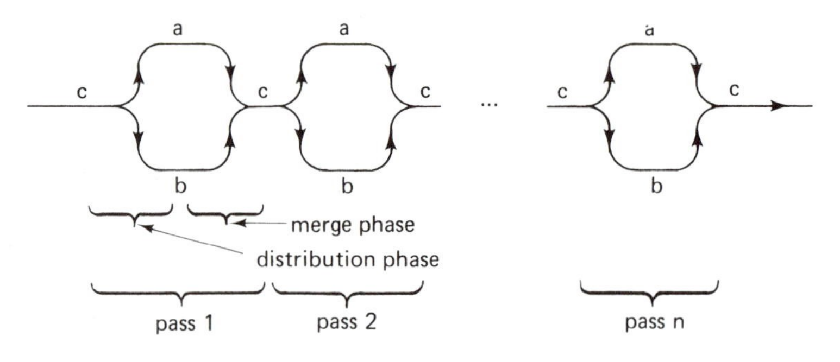

Fig. 2.13 Sort phases and sort passes.

```
17 31'05 59'13 41 43 67'11 23 29 47'03 07 71'02 19 57'37 61
05 17 31 59'11 13 23 29 41 43 47 67'02 03 07 19 57 71'37 61
05 11 13 17 23 29 31 41 43 47 59 67'02 03 07 19 37 57 61 71
02 03 05 07 11 13 17 19 23 29 31 37 41 43 47 57 59 61 67 71
```

Table 2.11. Example of a Natural Mergesort.

As an example, Table 2.11 shows the file c in its original state (line1) and after each pass (lines 2-4) in a natural merge sort involving 20 numbers. Note that only three passes are needed. The sort terminates as soon as the number of runs on c is 1. (We assume that there exists at least one non-empty run on the initial sequence.) We therefore let a variable L be used for counting the number of runs merged onto c. By making use of the type *Sequence* defined in Sect. 1.11, the program can be formulated as shown in (2.27).

```
VAR L: INTEGER;  a, b, c: Sequence

REPEAT Reset(a); Reset(b); Reset(c);
  distribute; (*c to a and b*)                                (2.26)
  Reset(a); Reset(b); Reset(c);
  L := 0; merge (*a and b into c*)
UNTIL L = 1
```

The two phases clearly emerge as two distinct statements. They are now to be refined, i.e., expressed in more detail. The refined description of *distribute* is given in (2.27) and that of *merge* in (2.28).

```
REPEAT copyrun(c, a);
  IF ~c.eof THEN copyrun(c, b) END                            (2.27)
UNTIL c.eof
```

```
REPEAT mergerun; L := L+1
UNTIL b.eof;                                                  (2.28)
IF ~a.eof THEN copyrun(a, c); L := L+1 END
```

This method of distribution supposedly results in either equal numbers of runs in both a and b, or in sequence a containing one run more than b. Since corresponding pairs of runs are merged, a leftover run may still be on file a, which simply has to be copied. The statements *merge* and *distribute* are formulated in terms of a refined statement *mergerun* and a subordinate procedure *copyrun* with obvious tasks. When attempting to do so, one runs into a serious difficulty: in order to determine the end of a run, two consecutive item keys must be compared. However, sequences are such that only a single element is immediately accessible. We evidently cannot avoid to *look ahead,* i.e to associate a buffer with every sequence. The buffer is to contain the first element of the sequence still to be read and constitutes something like a window sliding along the sequence.

Instead of programming this mechanism explicitly into our program, we prefer to define yet another level of abstraction. It is represented by new module called *Sequences* which replaces the module *FileSystem* for the clients and defines a new, appropriate sequence type, but itself refers to *FileSystem's* type *Sequence.* This new type will not only indicate the end of a sequence, but also the end of a run, and, of course, the first element of the remaining part of the sequence. The new type as well as its operators are presented by the following definition module.

```
DEFINITION MODULE Sequences;
  IMPORT FileSystem;                                          (2.29)
  TYPE item = INTEGER;

    Sequence =
      RECORD first: item;
           eor, eof: BOOLEAN;
```

```
      f: FileSystem.Sequence
    END ;
  PROCEDURE OpenSeq(VAR s: Sequence);
  PROCEDURE OpenRandomSeq(VAR s: Sequence; length, seed: INTEGER);
  PROCEDURE StartRead(VAR s: Sequence);
  PROCEDURE StartWrite(VAR s: Sequence);
  PROCEDURE copy(VAR x, y: Sequence);
  PROCEDURE CloseSeq(VAR s: Sequence);
  PROCEDURE ListSeq(VAR s: Sequence);
END Sequences.
```

A few additional explanations for the choice of the procedures are necessary. As we shall see, the sorting algorithms discussed here and later are based on copying elements from one sequence to another. A procedure *copy* therefore takes the place of separate read and write operations. After a file has been opened, the lookahead mechanism needs to be instructed as to whether the sequence is going to be written or read. In the latter case the first element needs to be read into the lookahead buffer *first*. Procedures *StartRead* and *StartWrite* therefore take over the role of *Reset* in *FileSystem*.

Merely for reasons of convenience two additional procedures are included. *OpenRandomSeq* takes the place of *OpenSeq*, if a sequence is to be initialized with numbers in random order, and the procedure *ListSeq* supposedly generates a listing of the specified sequence. These two procedures will serve to test the algorithms to be discussed below. The implementation of this concept is presented in the form of the following module. We note that in the procedure *copy* the field *first* of the destination sequence holds the value of the *last* item written. Also, the values of the fields *eof* and *eor* are defined as results of *copy* in analogy to *eof* having been defined as result of a read operation.

```
IMPLEMENTATION MODULE Sequences;
  FROM FileSystem IMPORT                                          (2.30)
    File, Open, ReadWord, WriteWord, Reset, Close;
  FROM InOut IMPORT WriteInt, WriteLn;

  PROCEDURE OpenSeq(VAR s: Sequence);
  BEGIN Open(s.f)
  END OpenSeq;

  PROCEDURE OpenRandomSeq(VAR s: Sequence; length, seed: INTEGER);
    VAR i: INTEGER;
  BEGIN Open(s.f);
    FOR i := 0 TO length-1 DO
      WriteWord(s.f, seed); seed := (31*seed) MOD 997 + 5
    END
  END OpenRandomSeq;

  PROCEDURE StartRead(VAR s: Sequence);
  BEGIN Reset(s.f); ReadWord(s.f, s.first); s.eof := s.f.eof
  END StartRead;

  PROCEDURE StartWrite(VAR s: Sequence);
  BEGIN Reset(s.f)
  END StartWrite;
```

```
  PROCEDURE copy(VAR x, y: Sequence);
  BEGIN y.first := x.first;
    WriteWord(y.f, y.first); ReadWord(x.f, x.first);
    x.eof := x.f.eof; x.eor := x.eof OR (x.first < y.first)
  END copy;

  PROCEDURE CloseSeq(VAR s: Sequence);
  BEGIN Close(s.f)
  END CloseSeq;

  PROCEDURE ListSeq(VAR s: Sequence);
    VAR i, L: CARDINAL;
  BEGIN Reset(s.f); i := 0; L := s.f.length;
    WHILE i < L DO
      WriteInt(INTEGER(s.f.a[i]), 6); i := i+1;
      IF i MOD 10 = 0 THEN WriteLn END
    END ;
    WriteLn
  END ListSeq;
END Sequences.
```

We now return to the process of successive refinement of the process of natural merging. Procedure *copyrun* and the statement *merge* are now conveniently expressible as shown in (2.31) and (2.32).

```
PROCEDURE copyrun(VAR x, y: Sequence);
BEGIN (*from x to y*)                                              (2.31)
  REPEAT copy(x, y) UNTIL x.eor
END copyrun

(*merge from a and b to c*)                                        (2.32)
REPEAT
  IF a.first < b.first THEN
    copy(a, c);
    IF a.eor THEN copyrun(b, c) END
  ELSE copy(b, c);
    IF b.eor THEN copyrun(a, c) END
  END
UNTIL a.eor OR b.eor
```

The comparison and selection process of keys in merging a run terminates as soon as one of the two runs is exhausted. After this, the other run (which is not exhausted yet) has to be transferred to the resulting run by merely copying its tail. This is done by a call of the procedure *copyrun*.

This should supposedly terminate the development of the natural merging sort procedure. Regrettably, the program is incorrect, as the very careful reader may have noticed. The program is incorrect in the sense that it does not sort properly in some cases. Consider, for example, the following sequence of input data:

```
03 02 05 11 07 13 19 17 23 31 29 37 43 41 47 59 57 61 71 67
```

By distributing consecutive runs alternately to a and b, we obtain

```
a =  03'07 13 19'29 37 43'57 61 71'
b =  02 05 11'17 23 31'41 47 59'67
```

These sequences are readily merged into a single run, whereafter the sort terminates successfully. The example, although it does not lead to an erroneous behaviour of the program, makes us aware that mere distribution of runs to serveral sequences may result in a number of output runs that is less than the number of input runs. This is because the first item of the i+2nd run may be larger than the last item of the i th run, thereby causing the two runs to merge automatically into a single run.

Although the procedure *distribute* supposedly outputs runs in equal numbers to the two sequences, the important consequence is that the actual number of resulting runs on a and b may differ significantly. Our merge procedure, however, only merges pairs of runs and terminates as soon as b is read, thereby losing the tail of one of the sequences. Consider the following input data that are sorted (and truncated) in two subsequent passes:

```
17 19 13 57 23 29 11 59 31 37 07 61 41 43 05 67 47 71 02 03
13 17 19 23 29 31 37 41 43 47 57 71 11 59
11 13 17 19 23 29 31 37 41 43 47 57 59 71
```

Table 2.12 Incorrect Result of Mergesort Program.

The example of this programming mistake is typical for many programming situations. The mistake is caused by an oversight of one of the possible consequences of a presumably simple operation. It is also typical in the sense that serval ways of correcting the mistake are open and that one of them has to be chosen. Often there exist two possibilities that differ in a very important, fundamental way:

1. We recognize that the operation of distribution is incorrectly programmed and does not satisfy the requirement that the number of runs differ by at most 1. We stick to the original scheme of operation and correct the faulty procedure accordingly.

2. We recognize that the correction of the faulty part involves far-reaching modifications, and we try to find ways in which other parts of the algorithm may be changed to accommodate the currently incorrect part.

In general, the first path seems to be the safer, cleaner one, the more honest way, providing a fair degree of immunity from later consequences of overlooked, intricate side effects. It is, therefore, the way toward a solution that is generally recommended.

It is to be pointed out, however, that the second possibility should sometimes not be entirely ignored. It is for this reason that we further elaborate on this example and illustrate a fix by modification of the merge procedure rather than the distribution procedure, which is primarily at fault.

This implies that we leave the distribution scheme untouched and renounce the condition that runs be equally distributed. This may result in a less than optimal performance. However, the worst-case performance remains unchanged, and moreover, the case of highly unequal distribution is statistically very unlikely. Efficiency considerations are therefore no serious argument against this solution.

If the condition of equal distribution of runs no longer exists, then the merge procedure has to be changed so that, after reaching the end of one file, the entire tail of the remaining file is copied instead of at most one run. This change is straightforward and is very simple in comparison with any change in the distribution scheme. (The reader is urged to convince himself of the truth of this claim). The revised version of the merge algorithm is included in the complete Program 2.14, in which procedures called only once have been directly substituted.

```
MODULE NaturalMerge;
  FROM Sequences IMPORT item, Sequence, OpenSeq, OpenRandomSeq,
     StartRead, StartWrite, copy, CloseSeq, ListSeq;

  VAR L: INTEGER;  (*no. of runs merged*)
    a, b, c: Sequence;
    ch: CHAR;

  PROCEDURE copyrun(VAR x, y: Sequence);
  BEGIN (*from x to y*)
    REPEAT copy(x, y) UNTIL x.eor
  END copyrun;

BEGIN OpenSeq(a); OpenSeq(b); OpenRandomSeq(c, 16, 531);
  ListSeq(c);
  REPEAT StartWrite(a); StartWrite(b); StartRead(c);
    REPEAT copyrun(c, a);
      IF ~c.eof THEN copyrun(c, b) END
    UNTIL c.eof;
    StartRead(a); StartRead(b); StartWrite(c);
    L := 0;
    REPEAT
      LOOP
        IF a.first < b.first THEN
          copy(a, c);
          IF a.eor THEN copyrun(b, c); EXIT END
        ELSE copy(b, c);
          IF b.eor THEN copyrun(a, c); EXIT END
        END
      END ;
      L := L+1
    UNTIL a.eof OR b.eof;
    WHILE ~a.eof DO copyrun(a, c); L := L+1 END ;
    WHILE ~b.eof DO copyrun(b, c); L := L+1 END
  UNTIL L = 1;
  ListSeq(c); CloseSeq(a); CloseSeq(b); CloseSeq(c)
END NaturalMerge.
```

Program 2.14 Natural Mergesort.

2.4.3 Balanced Multiway Merging

The effort involved in a sequential sort is proportional to the number of required passes since, by definition, every pass involves the copying of the entire set of data. One way to reduce this number is to distribute runs onto more than two sequences. Merging r runs that are equally distributed on N sequences results in a sequence of r/N runs. A second pass reduces their number to r/N^2, a third pass to r/N^3, and after k passes there are r/N^k runs left. The total number of passes required to sort n items by *N-way merging* is therefore $k = \lceil \log_N n \rceil$. Since each pass requires n copy operations, the total number of copy operations is in the worst case $M = n * \lceil \log_N n \rceil$

As the next programming exercise, we will develop a sort program based on multiway merging. In order to further contrast the program from the previous natural two-phase merging procedure, we shall formulate the multiway merge as a *single phase, balanced mergesort*. This implies that in each pass there are an equal number of input and output files onto which consecutive runs are alternately distributed. Using N sequences (N being even), the algorithm will therefore be based on N/2-way merging. Following the previously adopted strategy, we will not bother to detect the automatic merging of two consecutive runs distributed onto the same sequence. Consequently, we are forced to design the merge program whithout assuming strictly equal numbers of runs on the input sequences.

In this program we encounter for the first time a natural application of a data structure consisting of an array of files. As a matter of fact, it is surprising how strongly the following program differs from the previous one because of the change from two-way to multiway merging. The change is primarily a result of the circumstance that the merge process can no longer simply be terminated after one of the input runs is exhausted. Instead, a list of inputs that are still active, i.e., not yet exhausted, must be kept. Another complication stems from the need to switch the groups of input and output sequences after each pass.

We start out by defining, in addition to the two familiar types *item* and *sequence*, a type

seqno = [1 .. N] (2.33)

Obviously, sequence numbers are used to index the array of sequences of items. Let us then assume that the initial sequence of items is given as a variable

f0: Sequence (2.34)

and that for the sorting process n tapes are available, where n is even

f: ARRAY seqno OF Sequence (2.35)

A recommended technique of approaching the problem of tape switching is to introduce a tape index map. Instead of directly addressing a tape by its index i, it is addressed via a map t, i.e., instead of f_i we write f_{t_i}, where the map is defined as

t: ARRAY seqno OF seqno (2.36)

If initially $t_i = i$ for all i, then a switch consists in merely exchanging the pairs of map components $t_i \leftrightarrow t_{Nh+i}$ for all $i = 1 \ldots Nh$, where $Nh = N/2$. Consequently, we may always consider $f_{t_1}, \ldots, f_{t_{Nh}}$ as input sequences, and we may always consider $f_{t_{Nh+1}}, \ldots, f_{t_N}$ as output sequences. (Subsequently, we will simply call f_{t_j} *sequence j.*) The algorithm can now be formulated initially as follows:

```
MODULE BalancedMerge;
  VAR i, j: seqno;
    L: INTEGER; (*no. of runs distributed*)
    t: ARRAY seqno OF seqno;
BEGIN (*distribute initial runs to t[1] ... t[Nh]*)                (2.37)
  j := Nh; L := 0;
  REPEAT
    IF j < Nh THEN j := j+1 ELSE j := 1 END ;
    copy one run from f0 to sequence j;
    L := L+1
  UNTIL f0.eof;
  FOR i := 1 TO N DO t[i] := i END ;
  REPEAT (*merge from t[1] ... t[nh] to t[nh+1] ... t[n]*)
    reset input sequences;
    L := 0;
    j := Nh+1; (*j = index of output sequence*)
    REPEAT L := L+1;
      merge a run from inputs to t[j];
      IF j < N THEN j := j+1 ELSE j := Nh+1 END
    UNTIL all inputs exhausted;
    switch sequences
  UNTIL L = 1
  (*sorted sequence is t[1]*)
END BalancedMerge.
```

We now refine the statement for the initial distribution of runs; using the definition of sequence (2.26) and of copy (2.32), we replace *copy one run from f0 to sequence j* by

```
REPEAT copy(f0, f[j]) UNTIL f0.eor                                  (2.38)
```

Copying a run terminates when either the first item of the next run is encountered or when the end of the entire input file is reached.

In the actual sort algorithm, the following statements remain to be specified in more detail:

1. Reset input sequences
2. Merge a run from inputs to t_j
3. Switch sequences
4. All inputs exhausted

First, we must accurately identify the current input sequences. Notably, the number of *active* inputs may be less than N/2. In fact, there can be at most as many sources as there are runs; the sort terminates as soon as there is one single sequence left. This leaves open the possibility that at the initiation of the last sort pass there are fewer than N/2 runs. We therefore introduce a variable, say k1, to denote the actual number of inputs used. We incorporate the initialization of k1 in the statement *reset input sequences* as follows:

```
IF L < Nh THEN k1 := L ELSE k1 := Nh END ;
FOR i := 1 TO k1 DO StartRead(f[t[i]]) END
```

Naturally, statement (2) is to decrement k1 whenever an input source ceases. Hence, predicate (4) may easily be expressed by the relation k1 = 0. Statement (2), however, is more difficult to refine; it consists of the repeated selection of the least key among the available sources and its subsequent transport to the destination, i.e., the current output sequence. The process is further complicated by the necessity of determining the end of each run. The end of a run may be reached because (1) the subsequent key is less than the current key or (2) the end of the source is reached. In the latter case the source is eliminated by decrementing k1; in the former case the run is closed by excluding the sequence from further selection of items, but only until the creation of the current output run is completed. This makes it obvious that a second variable, say k2, is needed to denote the number of sources actually available for the selection of the next item. This value is initially set equal to k1 and is decremented whenever a run teminates because of condition (1).

Unfortunately, the introduction of k2 is not sufficient. We need to know not only the number of sequences, but also which sequences are still in actual use. An obvious solution is to use an array with Boolean components indicating the availability of the sequences. We choose, however, a different method that leads to a more efficient selection procedure which, after all, is the most frequently repeated part of the entire algorithm. Instead of using a Boolean array, a second tape map, say *ta*, is introduced. This map is used in place of t such that $ta_1 \dots ta_{k2}$ are the indices of the available sequences. Thus statement (2) can be formulated as follows:

```
k2 := k1;
REPEAT select the minimal key,
  let ta[mx] be the sequence number on which it occurs;
  copy(f[ta[mx]], f[t[j]]);                                  (2.39)
  IF f[ta[mx]].eof THEN eliminate tape
  ELSIF f[ta[mx]].eor THEN close run
  END
UNTIL k2 = 0
```

Since the number of sequences will be fairly small for any practical purpose, the selection algorithm to be specified in further detail in the next refinement step may as well be a straightforward linear search. The statement *eliminate tape* implies a decrease of k1 as well as k2 and also a reassignment of indices in the map *ta*. The statement *close run* merely decrements k2 and rearranges components of *ta* accordingly. The details are shown in Program 2.15, which is a last refinement of (2.37) through (2.39). The statement *switch sequences* is elaborated according to explanations given earlier.

```
MODULE BalancedMerge;
  FROM Sequences IMPORT item, Sequence, OpenSeq, OpenRandomSeq,
     StartRead, StartWrite, copy, CloseSeq, ListSeq;

  CONST N = 4; Nh = N DIV 2;
  TYPE seqno = [1 .. N];
  VAR i, j, mx, tx: seqno;
    L, k1, k2: CARDINAL;
    min, x: item;
    t, ta: ARRAY seqno OF seqno;
    f0: Sequence;
    f:  ARRAY seqno OF Sequence;

BEGIN OpenRandomSeq(f0, 100, 737); ListSeq(f0);
  FOR i := 1 TO N DO OpenSeq(f[i]) END ;
  (*distribute initial runs to t[1] ... t[Nh]*)
  FOR i := 1 TO Nh DO StartWrite(f[i]) END ;
  j := Nh; L := 0; StartRead(f0);
  REPEAT
    IF j < Nh THEN j := j+1 ELSE j := 1 END ;
    REPEAT copy(f0, f[j]) UNTIL f0.eor;
    L := L+1
  UNTIL f0.eof;
  FOR i := 1 TO N DO t[i] := i END ;
  REPEAT (*merge from t[1] ... t[nh] to t[nh+1] ... t[n]*)
    IF L < Nh THEN k1 := L ELSE k1 := Nh END ;
    FOR i := 1 TO k1 DO
      StartRead(f[t[i]]); ta[i] := t[i]
    END ;
    L := 0;     (*no. of runs merged*)
    j := Nh+1;  (*j = index of output sequence*)
    REPEAT (*merge a run from inputs to t[j]*)
      L := L+1; k2 := k1;
      REPEAT (*select the minimal key*)
        i := 1; mx := 1; min := f[ta[1]].first;
        WHILE i < k2 DO
          i := i+1; x := f[ta[i]].first;
          IF x < min THEN min := x; mx := i END
        END ;
        copy(f[ta[mx]], f[t[j]]);
        IF f[ta[mx]].eof THEN (*eliminate tape*)
          StartWrite(f[ta[mx]]); ta[mx] := ta[k2];
          ta[k2] := ta[k1]; k1 := k1-1; k2 := k2-1
        ELSIF f[ta[mx]].eor THEN (*close run*)
```

```
        tx := ta[mx]; ta[mx] := ta[k2]; ta[k2] := tx; k2 := k2-1
      END
    UNTIL k2 = 0;
    IF j < N THEN j := j+1 ELSE j := Nh+1 END
  UNTIL k1 = 0;
  FOR i := 1 TO Nh DO
    tx := t[i]; t[i] := t[i+Nh]; t[i+Nh] := tx;
  END
UNTIL L = 1;
ListSeq(f[t[1]])
(*sorted sequence is t[1]*)
END BalancedMerge.
```

Program 2.15 Balanced Mergesort

2.4.4. Polyphase Sort

We have now discussed the necessary techniques and have acquired the proper background to investigate and program yet another sorting algorithm whose performance is superior to the balanced sort. We have seen that balanced merging eliminates the pure copying operations necessary when the distribution and the merging operations are united into a single phase. The question arises whether or not the given sequences could be processed even more efficiently. This is indeed the case; the key to this next improvement lies in abandoning the rigid notion of strict passes, i.e., to use the sequences in a more sophisticated way than by always having N/2 sources and as many destinations and exchanging sources and destinations at the end of each distinct pass. Instead, the notion of a pass becomes diffuse. The method was invented by R.L. Gilstad [2-3] and called *Polyphase Sort.*

It is first illustrated by an example using three sequences. At any time, items are merged from two sources into a third sequence variable. Whenever one of the source sequences is exhausted, it immediately becomes the destination of the merge operations of data from the non-exhausted source and the previous destination sequence.

As we know that n runs on each input are transformed into n runs on the output, we need to list only the number of runs present on each sequence (instead of specifying actual keys). In Fig. 2.14 we assume that initially the two input sequences f_1 and f_2 contain 13 and 8 runs, respectively. Thus, in the first pass 8 runs are merged from f_1 and f_2 to f_3, in the second pass the remaining 5 runs are merged from f_3 and f_1 onto f_2, etc. In the end, f_1 is the sorted sequence.

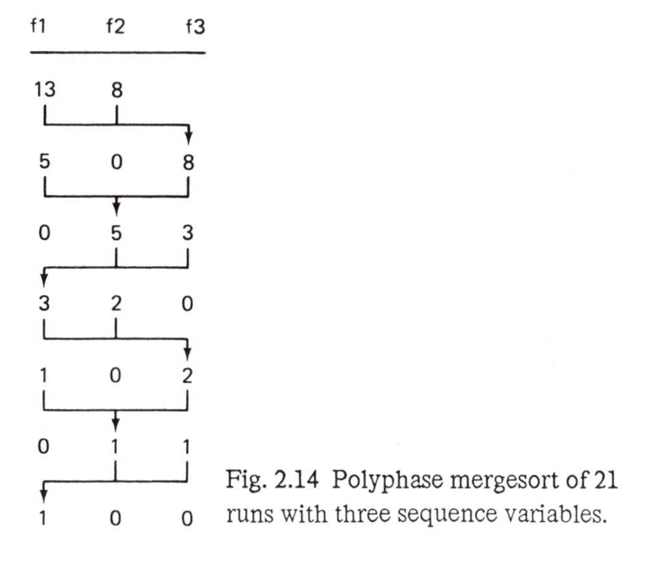

Fig. 2.14 Polyphase mergesort of 21 runs with three sequence variables.

A second example shows the Polyphase method with 6 sequences. Let there initially be

16 runs on f_1, 15 on f_2, 14 on f_3, 12 on f_4, and 8 on f_5. In the first partial pass, 8 runs are merged onto f_6; In the end, f_2 contains the sorted set of items (see Fig. 2.15).

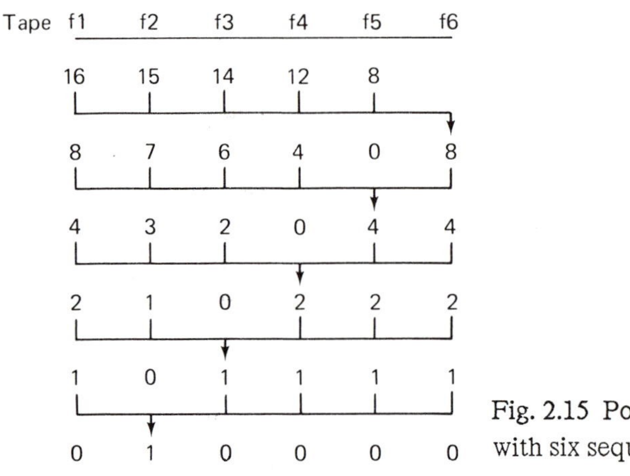

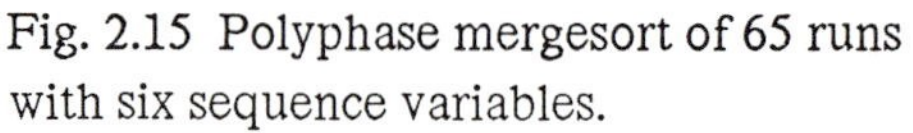
Fig. 2.15 Polyphase mergesort of 65 runs with six sequence variables.

Polyphase is more efficient than balanced merge because, given N sequences, it always operates with an N-1-way merge instead of an N/2-way merge. As the number of required passes is approximately $\log_N n$, n being the number of items to be sorted and N being the degree of the merge operations, Polyphase promises a significant improvement over balanced merging.

Of course, the distribution of initial runs was carefully chosen in the above examples. In order to find out which initial distributions of runs lead to a proper functioning, we work backward, starting with the final distribution (last line in Fig. 2.15). Rewriting the tables of the two examples and rotating each row by one position with respect to the prior row yields Tables 2.13 and 2.14 for six passes and for three and six sequences, respectively.

L	$a_1^{(L)}$	$a_2^{(L)}$	Sum $a_i^{(L)}$
0	1	0	1
1	1	1	2
2	2	1	3
3	3	2	5
4	5	3	8
5	8	5	13
6	13	8	21

Table 2.13 Perfect distribution of runs on two sequences.

L	$a_1^{(L)}$	$a_2^{(L)}$	$a_3^{(L)}$	$a_4^{(L)}$	$a_5^{(L)}$	Sum $a_i^{(L)}$
0	1	0	0	0	0	1
1	1	1	1	1	1	5

2	2	2	2	2	1	9
3	4	4	4	3	2	17
4	8	8	7	6	4	33
5	16	15	14	12	8	65

Table 2.14 Perfect distribution of runs on five sequences.

From Table 2.13 we can deduce for $L > 0$ the relations

$$\begin{aligned} a_2^{(L+1)} &= a_1^{(L)} \\ a_1^{(L+1)} &= a_1^{(L)} + a_2^{(L)} \end{aligned} \tag{2.40}$$

and $a_1^{(0)} = 1$, $a_2^{(0)} = 0$. Defining $f_{i+1} = a_1^{(i)}$, we obtain for $i > 0$

$$f_{i+1} = f_i + f_{i-1},\ f_1 = 1,\ f_0 = 0 \tag{2.41}$$

These are the recursive rules (or recurrence relations) defining the so-called *Fibonacci numbers:*

$$f = 0, 1, 1, 2, 3, 5, 8, 13, 21, 34, 55, \ldots$$

Each Fibonacci number is the sum of its two predecessors. As a consequence, the numbers of initial runs on the two input sequences must be two consecutive Fibonacci numbers in order to make Polyphase work properly with three sequences.

How about the second example (Table 2.14) with six sequences? The formation rules are easily derived as

$$\begin{aligned} a_5^{(L+1)} &= a_1^{(L)} \\ a_4^{(L+1)} &= a_1^{(L)} + a_5^{(L)} = a_1^{(L)} + a_1^{(L-1)} \\ a_3^{(L+1)} &= a_1^{(L)} + a_4^{(L)} = a_1^{(L)} + a_1^{(L-1)} + a_1^{(L-2)} \\ a_2^{(L+1)} &= a_1^{(L)} + a_3^{(L)} = a_1^{(L)} + a_1^{(L-1)} + a_1^{(L-2)} + a_1^{(L-3)} \\ a_1^{(L+1)} &= a_1^{(L)} + a_2^{(L)} = a_1^{(L)} + a_1^{(L-1)} + a_1^{(L-2)} + a_1^{(L-3)} + a_1^{(L-4)} \end{aligned} \tag{2.42}$$

Substituting f_i for $a_1^{(i)}$ yields

$$\begin{aligned} f_{i+1} &= f_i + f_{i-1} + f_{i-2} + f_{i-3} + f_{i-4} \quad \text{for } i \geq 4 \\ f_4 &= 1 \\ f_i &= 0 \quad \text{for } i < 4 \end{aligned} \tag{2.43}$$

These numbers are the so-called Fibonacci numbers of order 4. In general, the *Fibonacci numbers of order p* are defined as follows:

$$\begin{aligned} f_{i+1}^{(p)} &= f_i^{(p)} + f_{i-1}^{(p)} + \ldots + f_{i-p}^{(p)} \quad \text{for } i \geq p \\ f_p^{(p)} &= 1 \\ f_i^{(p)} &= 0 \quad \text{for } 0 \leq i < p \end{aligned} \tag{2.44}$$

Note that the ordinary Fibonacci numbers are those of order 1.

We have now seen that the initial numbers of runs for a perfect Polyphase Sort with N sequences are the sums of any N-1, N-2, ... , 1 (see Table 2.15) consecutive Fibonacci numbers of order N-2. This apparently implies that this method is only applicable to inputs whose number of runs is the sum of N-1 such Fibonacci sums. The important question thus arises: What is to be done when the number of initial runs is not such an ideal sum? The answer is simple (and typical for such situations): we simulate the existence of hypothetical empty runs, such that the sum of real and hypothetical runs is a perfect sum. The empty runs are called *dymmy runs*.

But this is not really a satisfactory answer because it immediately raises the further and more difficult question: How do we recognize dummy runs during merging? Before answering this question we must first investigate the prior problem of initial run distribution and decide upon a rule for the distribution of actual and dummy runs onto the N-1 tapes.

2	3	5	7	9	11	13
3	5	9	13	17	21	25
4	8	17	25	33	41	49
5	13	31	49	65	81	97
6	21	57	94	129	161	193
7	34	105	181	253	321	385
8	55	193	349	497	636	769
9	89	355	673	977	1261	1531
10	144	653	1297	1921	2501	3049
11	233	1201	2500	3777	4961	6073
12	377	2209	4819	7425	9841	12097
13	610	4063	9289	14597	19521	24097
14	987	7473	17905	28697	38721	48001

Table 2.15 Numbers of runs allowing for perfect distribution.

In order to find an appropriate rule for distribution, however, we must know how actual and dummy runs are merged. Clearly, the selection of a dummy run from sequence i means precisely that sequence i is ignored during this merge. resulting in a merge from fewer than N-1 sources. Merging of a dummy run from all N-1 sources implies no actual merge operation, but instead the recording of the resulting dummy run on the output sequence. From this we conclude that dummy runs should be distributed to the n-1 sequences as uniformaly as possible, since we are interested in active merges from as many sources as possible.

Let us forget dummy runs for a moment and consider the problem of distributing an *unknown* number of runs onto N-1 sequences. It is plain that the Fibonacci numbers of order N-2 specifying the desired numbers of runs on each source can be generated while the distribution progresses. Assuming, for example, N = 6 and referring to Table 2.14, we start by distributing runs as indicated by the row with index L = 1 (1, 1, 1, 1, 1); if there are more runs available, we proceed to the second row (2, 2, 2, 2, 1); if the source is still not exhausted, the distribution proceeds according to the third row (4, 4, 4, 3, 2), and so on. We shall call the row index *level*. Evidently, the larger the number of runs, the higher is the level of Fibonacci numbers which, incidentally, is equal to the number of merge passes or switchings necessary for the subsequent sort. The distribution algorithm can now be formulated in a first version as follows:

1. Let the distribution goal be the Fibonacci numbers of order N-2, level 1.
2. Distribute according to the set goal.
3. If the goal is reached, compute the next level of Fibonacci numbers; the difference between them and those on the former level constitutes the new distribution goal. Return to step 2. If the goal cannot be reached because the source is exhausted, terminate the distribution process.

The rules for calculating the next level of Fibonacci numbers are contained in their definition (2.44). We can thus concentrate our attention on step 2, where, with a given goal, the subsequent runs are to be distributed one after the other onto the N-1 output sequences. It is here where the dummy runs have to reappear in our considerations.

Let us assume that when raising the level, we record the next goal by the differences d_i for $i = 1 ... N-1$, where d_i denotes the number of runs to be put onto sequence i in this step. We can now assume that we immediately put d_i dummy runs onto sequence i and then regard the subsequent distribution as the replacement of dummy runs by actual runs, each time recording a replacement by subtracting 1 from the count d_i. Thus, the d_i indicates the number of dummy runs on sequence i when the source becomes empty.

It is not known which algorithm yields the optimal distribution, but the following has proved to be a very good method. It is called *horizontal distribution* (cf. Knuth, vol 3. p. 270), a term that can be understood by imagining the runs as being piled up in the form of silos, as shown in Fig. 2.16 for N = 6, level 5 (cf. Table 2.14).

In order to reach an equal distribution of remaining dummy runs as quickly as possible, their replacement by actual runs reduces the size of the piles by picking off dummy runs on horizontal levels proceeding from left to right. In this way, the runs are distributed onto the sequences as indicated by their numbers as shown in Fig. 2.16.

8	1				
7	2	3	4		
6	5	6	7	8	
5	9	10	11	12	
4	13	14	15	16	17
3	18	19	20	21	22
2	23	24	25	26	27
1	28	29	30	31	32

Fig. 2.16 Horizontal distribution of runs.

We are now in a position to describe the algorithm in the form of a procedure called *select*, which is activated each time a run has been copied and a new source is selected for the next run. We assume the existence of a variable j denoting the index of the current destination sequence. a_i and d_i denote the ideal and dummy distribution numbers for sequence i.

```
j:      seqno;
a, d:   ARRAY seqno OF INTEGER;
level:  INTEGER
```

These variables are initialized with the following values:

$a_i = 1,\ d_i = 1 \quad$ for i = 1 ... N-1
$a_N = 0, d_N = 0 \quad$ (dummy)
$j = 1, \quad level = 1$

Note that *select* is to compute the next row of Table 2.14, i.e., the values $a_1^{(L)} \dots a_{N-1}^{(L)}$ each time that the level is increased. The next goal, i.e., the differences $d_i = a_i^{(L)} - a_i^{(L-1)}$ are also computed at that time. The indicated algorithm relies on the fact that the resulting d_i decrease with increasing index (descending stair in Fig. 2.16). Note that the exception is the transition from level 0 to level 1; this algorithm must therefore be used starting at level 1. *Select* ends by decrementing d_j by 1; this operation stands for the replacement of a dummy run on sequence j by an actual run.

Assuming the availability of a routine to copy a run from the source f_0 onto f_j, we can formulate the initial distribution phase as follows (always assuming that the source contains at least one run):

```
REPEAT select; copyrun
UNTIL f0.eof                                                    (2.47)
```

Here, however, we must pause for a moment to recall the effect encountered in distributing runs in the previously discussed natural merge algorithm: the fact that two runs consecutively arriving at the same destination may turn out to constitute a single run causes the assumed numbers of runs to be incorrect. By devising the sort algorithm such that its correctness does not depend on the number of runs, this side effect can safely be ignored. In the Polyphase Sort, however, we are particularly concerned about keeping track of the exact number of runs on each file. Consequently, we cannot afford to overlook the effect of such a coincidental merge. An additional complication of the distribution algorithm therefore cannot be avoided. It becomes necessary to retain the keys of the last item of the last run on each sequence. Fortunately, out implementation of Sequences does exactly this. In the case of output sequences, *f.first* represents the item last written. A next attempt to describe the distribution algorithm could therefore be

```
REPEAT select;                                                  (2.48)
  IF f[j].first <= f0.first THEN continue old run END ;
  copyrun
UNTIL f0.eof
```

The obvious mistake lies in forgetting that f[j].first has only obtained a value after copying the first run. A correct solution must therefore first distribute one run onto each of the N-1 destination sequences without inspection of f[j].first. The remaining runs are distributed according to (2.49).

```
WHILE ~f0.eof DO                                              (2.49)
  select;
  IF f[j].first <= f0.first THEN
    copyrun;
    IF f0.eof THEN d[j] := d[j] + 1 ELSE copyrun END
  ELSE copyrun
  END
END
```

Now we are finally in a position to tackle the main polyphase merge sort algorithm. Its principal structure is similar to the main part of the N-way merge program: an outer loop whose body merges runs until the sources are exhausted, an inner loop whose body merges a single run from each source, and an innermost loop whose body selects the initial key and transmits the involved item to the target file. The principal differences are the following:

1. Instead of N/2, there is only one output sequence in each pass.
2. Instead of switching N/2 input and N/2 output sequences after each pass, the sequences are rotated. This is achieved by using a sequence index map t.
3. The number of input sequences varies from run to run; at the start of each run, it is determined from the counts d_i of dummy runs. If $d_i > 0$ for all i, then N-1 dummy runs are pseudo-merged into one dummy run by merely incrementing the count d_N of the output sequence. Otherwise, one run is merged from all sources with $d_i = 0$, and d_i is decremented for all other sequences, indicating that one dummy run was taken off. We denote the number of input sequences involved in a merge by k.
4. It is impossible to derive termination of a phase by the end-of status of the N-1'st sequence, because more merges might be necessary involving dummy runs from that source. Instead, the theoretically necessary number of runs is determined from the coefficients a_i. The coefficients a_i were computed during the distribution phase; they can now be recomputed backward.

The main part of the Polyphase Sort can now be formulated according to these rules, assuming that all N-1 sequences with initial runs are set to be read, and that the tape map is initially set to $t_i = i$.

```
REPEAT (*merge from t[1] ... t[N-1] to t[N]*)                 (2.50)
  z := a[N-1]; d[N] := 0; StartWrite(f[t[N]]);
  REPEAT k := 0; (*merge one run*)
    (*determine no. of active input sequences*)
    FOR i := 1 TO N-1 DO
      IF d[i] > 0 THEN d[i] := d[i] - 1
      ELSE k := k+1; ta[k] := t[i]
      END
    END ;
```

```
      IF k = 0 THEN d[N] := d[N] + 1
      ELSE merge one real run from t[1] ... t[k] to t[N]
      END ;
      z := z-1
    UNTIL z = 0;
    StartRead(f[t[N]]);
    rotate sequences in map t; compute a[i] for next level;
    StartWrite(f[t[N]]); level := level - 1
  UNTIL level = 0
  (*sorted output is t[1]*)
```

The actual merge operation is almost identical with that of the N-way merge sort, the only difference being that the sequence elimination algorithm is somewhat simpler. The rotation of the sequence index map and the corresponding counts d_i (and the down-level recomputation of the coefficients a_i) is straightforward and can be inspected in detail from Program 2.16, which represents the Polyphase algorithm in its entirety.

```
MODULE Polyphase;
  FROM Sequences IMPORT item, Sequence, OpenSeq, OpenRandomSeq,
     StartRead, StartWrite, copy, CloseSeq, ListSeq;

  CONST N = 6;
  TYPE seqno = [1 .. N];


  VAR i, j, mx, tn: seqno;
    k, dn, z, level: CARDINAL;
    x, min: item;
    a, d:  ARRAY seqno OF CARDINAL;
    t, ta: ARRAY seqno OF seqno;
    f0:   Sequence;
    f:    ARRAY seqno OF Sequence;

  PROCEDURE select;
    VAR i, z: CARDINAL;
  BEGIN
    IF d[j] < d[j+1] THEN j := j+1
    ELSE
      IF d[j] = 0 THEN
        level := level + 1; z := a[1];
        FOR i := 1 TO N-1 DO
          d[i] := z + a[i+1] - a[i]; a[i] := z + a[i+1]
        END
      END ;
      j := 1
    END ;
    d[j] := d[j] - 1
  END select;
```

```
  PROCEDURE copyrun; (*from f0 to f[j]*)
  BEGIN
    REPEAT copy(f0, f[j]) UNTIL f0.eor
  END copyrun;

BEGIN OpenRandomSeq(f0, 100, 561); ListSeq(f0);
  FOR i := 1 TO N DO OpenSeq(f[i]) END ;
  (*distribute initial runs*)
  FOR i := 1 TO N-1 DO
    a[i] := 1; d[i] := 1; StartWrite(f[i])
  END ;
  level := 1; j := 1; a[N] := 0; d[N] := 0; StartRead(f0);
  REPEAT select; copyrun
  UNTIL f0.eof OR (j = N-1);
  WHILE ~f0.eof DO
    select; (*f[j].first = last item written on f[j]*)
    IF f[j].first <= f0.first THEN
      copyrun;
      IF f0.eof THEN d[j] := d[j] + 1 ELSE copyrun END
    ELSE copyrun
    END
  END ;

  FOR i := 1 TO N-1 DO t[i] := i; StartRead(f[i]) END ;
  t[N] := N;
  REPEAT (*merge from t[1] ... t[N-1] to t[N]*)
    z := a[N-1]; d[N] := 0; StartWrite(f[t[N]]);
    REPEAT k := 0; (*merge one run*)
      FOR i := 1 TO N-1 DO
        IF d[i] > 0 THEN d[i] := d[i] - 1
        ELSE k := k+1; ta[k] := t[i]
        END
      END ;
      IF k = 0 THEN d[N] := d[N] + 1
      ELSE (*merge one real run from t[1] ... t[k] to t[N]*)
        REPEAT i := 1; mx := 1; min := f[ta[1]].first;
          WHILE i < k DO
            i := i+1; x := f[ta[i]].first;
            IF x < min THEN min := x; mx := i END
          END ;
          copy(f[ta[mx]], f[t[N]]);
          IF f[ta[mx]].eor THEN (*drop this source*)
            ta[mx] := ta[k]; k := k-1
          END
        UNTIL k = 0
      END ;
```

```
      z := z-1
    UNTIL z = 0;
    StartRead(f[t[N]]);  (*rotate sequences*)
    tn := t[N]; dn := d[N]; z := a[N-1];
    FOR i := N TO 2 BY -1 DO
      t[i] := t[i-1]; d[i] := d[i-1]; a[i] := a[i-1] - z
    END ;
    t[1] := tn; d[1] := dn; a[1] := z;
    StartWrite(f[t[N]]); level := level - 1
  UNTIL level = 0 ;
  ListSeq(f[t[1]]);
  FOR i := 1 TO N DO CloseSeq(f[i]) END
END Polyphase.
```

Program 2.16. Polyphase sort.

2.4.5. Distribution of Initial Runs

We were led to the sophisticated sequential sorting programs, because the simpler methods operating on arrays rely on the availability of a random access store sufficiently large to hold the entire set of data to be sorted. Very often such a store is unavailable; instead, sufficiently large sequential storage devices such as tapes or disks must be used. We know that the sequential sorting methods developed so far need practically no primary store whatsoever, except for the file buffers and, of course, the program itself. However, it is a fact that even small computers include a random access, primary store that is almost always larger than what is needed by the programs developed here. Failing to make optimal use of it cannot be justified.

The solution lies in combining array and sequence sorting techniques. In particular, an adapted array sort may be used in the distribution phase of initial runs with the effect that these runs do already have a length L of approximately the size of the available primary data store. It is plain that in the subsequent merge passes no additional array sorts could improve the performance because the runs involved are steadily growing in length, and thus they always remain larger than the available main store. As a result, we may fortunately concentrate our attention on improving the algorithm that generates initial runs.

Naturally, we immediately concentrate our search on the logarithmic array sorting methods. The most suitable of them is the tree sort or Heapsort method (see Sect. 2.2.5). The heap may be regarded as a tunnel through which all items must pass, some quicker and some more slowly. The least key is readily picked off the top of the heap, and its replacement is a very efficient process. The action of funnelling a component from the input sequence f0 through a full heap tunnel H onto an output sequence f[j] may be described simply as follows:

```
WriteWord(f[j], H[1]); ReadWord(f0, H[1]); sift(1, n)          (2.51)
```

Sift is the process described in Sect. 2.2.5 for sifting the newly inserted component H_1

down into its proper place. Note that H_1 is the least item on the heap. An example is shown in Fig. 2.17. The program eventually becomes considerably more complex for the following reasons:

1. The heap H is initially empty and must first be filled.
2. Toward the end, the heap is only partially filled, and it ultimately becomes empty.
3. We must keep track of the beginning of new runs in order to change the output index j at the right time.

State before a transfer:

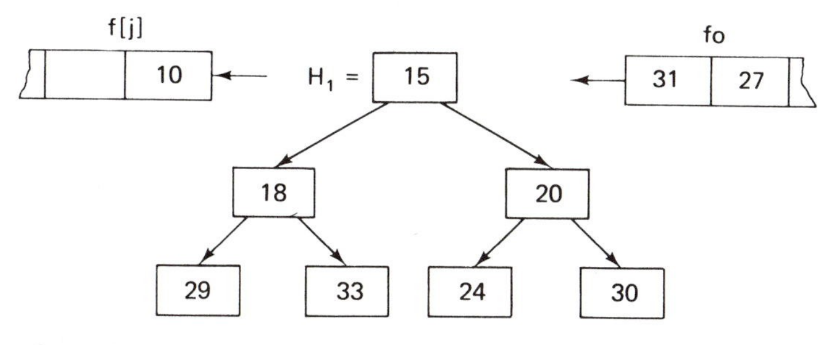

State after the next transfer:

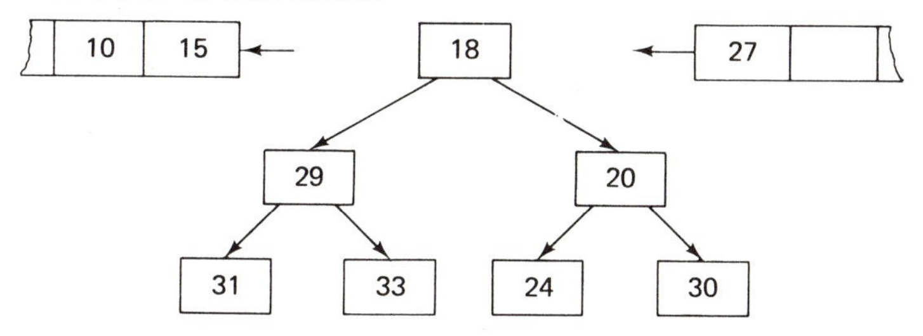

Fig. 2.17 Sifting a key through a heap.

Before proceeding, let us formally declare the variables that are evidently involved in the process:

```
VAR f0: Sequence;
  f: ARRAY seqno OF Sequence;
  H: ARRAY [1 .. m] OF item;                    (2.52)
  L, R: INTEGER
```

m is the size of the heap h. We use the constant mh to denote m/2; L and R are indices delimiting the heap. The funnelling process can then be divided into five distinct parts.

1. Read the first mh items from f_0 and put them into the upper half of the heap where no ordering among the keys is prescribed.
2. Read another mh items and put them into the lower half of the heap, sifting each item into its appropriate position (build heap).

3. Set L to m and repeat the following step for all remaining items on f_0: Feed H_1 to the appropriate output sequence. If its key is less or equal to the key of the next item on the input sequence, then this next item belongs to the same run and can be sifted into its proper position. Otherwise, reduce the size of the heap and place the new item into a second, upper heap that is built up to contain the next run. We indicate the borderline between the two heaps with the index L. Thus, the lower (current) heap consists of the items $H_1 \dots H_L$, the upper (next) heap of $H_{L+1} \dots H_m$. If L = 0, then switch the output and reset L to m.
4. Now the source is exhausted. First, set R to m; then flush the lower part terminating the current run, and at the same time build up the upper part and gradually relocate it into positions $H_{L+1} \dots H_R$.
5. The last run is generated from the remaining items in the heap.

We are now in a position to describe the five stages in detail as a complete program, calling a procedure *select* whenever the end of a run is detected and some action to alter the index of the output sequence has to be invoked. In Program 2.17 a dummy routine is used instead; it merely counts the number of runs generated. All elements are written onto sequence f_1.

If we now try to integrate this program with, for instance, Polyphase Sort, we encounter a serious difficulty. It arises from the following circumstances: The sort program consists in its initial part of a fairly complicated routine for switching between sequence variables, and relies on the availability of a procedure *copyrun* that delivers exactly one run to the selected destination. The *Heapsort* program, on the other hand, is a complex routine relying on the availability of a closed procedure *select* which simply selects a new destination. There would be no problem, if in one (or both) of the programs the required procedure would be called at a single place only; but instead, they are called at several places in both programs.

This situation is best reflected by the use of a so-called *coroutine;* it is suitable in those cases in which several processes coexist. The most typical representative is the combination of a process that produces a stream of information in distinct entities and a process that consumes this stream. This producer-consumer relationship can be expressed in terms of two coroutines; one of them may well be the main program itself. The coroutine may be considered as a process that contains one or more breakpoints. If such a breakpoint is encountered, then control returns to the program that had activated the coroutine. Whenever the coroutine is called again, execution is resumed at that breakpoint. In our example, we might consider Polyphase Sort as the main program, calling upon *copyrun*, which is formulated as a coroutine. It consists of the main body of Program 2.17 in which each call of *select* now represents a breakpoint. The test for end of file would then have to be replaced systematically by a test of whether or not the coroutine had reached its endpoint.

```
MODULE Distribute; (*by HeapSort*)
  FROM InOut IMPORT WriteInt, WriteLn;
  FROM FS IMPORT File, Open, ReadWord, WriteWord, Reset, Close;
```

```
  CONST m = 16; mh = m DIV 2;  (*heap size*)
  TYPE item = INTEGER;

  VAR L, R: CARDINAL;
    count: CARDINAL;
    x: item;
    H: ARRAY [1 .. m] OF item;  (*heap*)
    f0, f1: File;

  PROCEDURE select;
  BEGIN count := count + 1  (*dummy*)
  END select;

  PROCEDURE sift(L, R: CARDINAL);
    VAR i, j: CARDINAL;  x: item;
  BEGIN i := L; j := 2*L; x := H[L];
    IF (j < R) & (H[j] > H[j+1]) THEN j := j+1 END ;
    WHILE (j <= R) & (x > H[j]) DO
      H[i] := H[j]; i := j; j := 2*j;
      IF (j < R) & (H[j] > H[j+1]) THEN j := j+1 END
    END ;
    H[i] := x
  END sift;

  PROCEDURE OpenRandomSeq(VAR s: File; length, seed: INTEGER);
    VAR i: INTEGER;
  BEGIN Open(s);
    FOR i := 0 TO length-1 DO
      WriteWord(s, seed); seed := (31*seed) MOD 997 + 5
    END
  END OpenRandomSeq;

  PROCEDURE List(VAR s: File);
    VAR i, L: CARDINAL;
  BEGIN Reset(s); i := 0; L := s.length;
    WHILE i < L DO
      WriteInt(INTEGER(s.a[i]), 6); i := i+1;
      IF i MOD 10 = 0 THEN WriteLn END
    END ;
    WriteLn
  END List;

BEGIN count := 0; OpenRandomSeq(f0, 200, 991); List(f0);
  Open(f1); Reset(f0);
  select;
(*step 1: fill upper half of heap*)
  L := m;
  REPEAT ReadWord(f0, H[L]); L := L-1
```

```
    UNTIL L = mh;
  (*step 2: fill lower half of heap*)
    REPEAT ReadWord(f0, H[L]); sift(L,m); L := L-1
    UNTIL L = 0;
  (*step 3: pass elements through heap*)
    L := m; ReadWord(f0, x);
    WHILE ~f0.eof DO
      WriteWord(f1, H[1]);
      IF H[1] <= x THEN
        (*x belongs to same run*) H[1] := x; sift(1,L)
      ELSE (*start next run*)
        H[1] := H[L]; sift(1, L-1); H[L] := x;
        IF L <= mh THEN sift(L,m) END ;
        L := L-1;
        IF L = 0 THEN
          (*heap full; start new run*) L := m; select
        END
      END ;
      ReadWord(f0, x)
    END ;
  (*step 4: flush lower half of heap*)
    R := m;
    REPEAT WriteWord(f1, H[1]);
      H[1] := H[L]; sift(1, L-1); H[L] := H[R]; R := R-1;
      IF L <= mh THEN sift(L,R) END ;
      L := L-1
    UNTIL L = 0;
  (*step 5: flush upper half of heap*)
    select;
    WHILE R > 0 DO
      WriteWord(f1, H[1]); H[1] := H[R]; R := R-1; sift(1,R)
    END ;
    List(f1);
    Close(f0); Close(f1)
  END Distribute.
```

Program 2.17. Distribution of Initial Runs Through a Heap.

Analysis and conclusions. What performance can be expected from a Polyphase Sort with initial distribution of runs by a Heapsort? We first discuss the improvement to be expected by introducing the heap.

In a sequence with randomly distributed keys the expected average length of runs is 2. What is this length after the sequence has been funnelled through a heap of size m? One is inclined to say m, but, fortunately, the actual result of probabilistic analysis is much better, namely 2m (see Knuth, vol. 3, p. 254). Therefore, the expected improvement factor is m.

An estimate of the performance of Polyphase can be gathered from Table 2.15, indicating the maximal number of initial runs that can be sorted in a given number of partial passes (levels) with a given number N of sequences. As an example, with six sequences and a heap of size m = 100, a file with up to 165,680,100 initial runs can be sorted within 10 partial passes. This is a remarkable performance.

Reviewing again the combination of Polyphase Sort and Heapsort, one cannot help but be amazed at the complexity of this program. After all, it performs the same easily defined task of permuting a set of items as is done by any of the short programs based on the straight array sorting principles. The moral of the entire chapter may be taken as an exhibition of the following:

1. The intimate connection between algorithm and underlying data structure, and in particular the influence of the latter on the former.
2. The sophistication by which the performance of a program can be improved, even when the available structure for its data (sequence instead of array) is rather ill-suited for the task.

EXERCISES

2.1. Which of the algorithms given by Programs 2.1 through 2.6, 2.8, 2.10, and 2.13 are stable sorting methods?

2.2. Would Program 2.2 still work correctly if $L < R$ were replaced by $L \leq R$ in the while clause? Would it still be correct if the statement L := m+1 were simplified to L := m and R := m? If not, find sets of values $a_1 \dots a_n$ upon which the altered program would fail.

2.3. Program and measure the execution time of the three straight sorting methods on your computer, and find coefficients by which the factors C and M have to be multiplied to yield real time estimates.

2.4. Specifty invariants for the repetitions in the three straight sorting algorithms.

2.5. Consider the following "obvious" version of the Parition Program 2.9 and find sets of values $a_i \dots a_n$ for which this version fails:

```
i := 1; j := n; x := a[n DIV 2];
REPEAT
  WHILE a[i] < x DO i := i+1 END ;
  WHILE x < a[j] DO j := j-1 END ;
  w := a[i]; a[i] := a[j]; a[j] := w
UNTIL i > j
```

2.6. Write a program that combines the Quicksort and Bubblesort algorithms as follows: Use Quicksort to obtain (unsorted) partitions of length m ($1 \leq m \leq n$); then use Bubblesort to complete the task. Note that the latter may sweep over the entire array of n elements, hence, minimizing the bookkeeping effort. Find that value of m which minimizes the total sort time. Note: Clearly, the optimum value of m will be quite small. It may therefore pay to let the Bubblesort sweep exactly m-1 times over the array instead of including a last pass establishing the fact that no further exchange is necessary.

2.7. Perform the same experiment as in Exercise 2.6 with a straight selection sort instead of a Bubblesort. Naturally, the selection sort cannot sweep over the whole array; therefore, the expected amount of index handling is somewhat greater.

2.8. Write a recursive Quicksort algorithm according to the recipe that the sorting of the shorter partition should be tackled before the sorting of the longer partition. Perform the former task by an iterative statement, the latter by a recursive call. (Hence, your sort procedure will contain one recursive call instead of two in Program 2.10 and none in Program 2.11.)

2.9. Find a permutation of the keys 1, 2, ... , n for which Quicksort displays its worst (best) behavior (n = 5, 6, 8).

2.10. Construct a natural merge program similar to the straight merge Program 2.13, operating on a double length array from both ends inward; compare its performance with that of Program 2.13.

2.11. Note that in a (two-way) natural merge we do not blindly select the least value among the available keys. Instead, upon encountering the end of a run, the tail of the other run is simply copied onto the output sequence. For example, merging of

2, 4, 5, 1, 2, ...
3, 6, 8, 9, 7, ...

results in the sequence

2, 3, 4, 5, 6, 8, 9, 1, 2, ...

instead of

2, 3, 4, 5, 1, 2, 6, 8, 9, ...

which seems to be better ordered. What is the reason for this strategy?

2.12. A sorting method similar to the Polyphase is the so-called *Cascade* merge sort [2.1 and 2.9]. It uses a different merge pattern. Given, for instance, six sequences T1, ... ,T6, the cascade merge, also starting with a "perfect distribution" of runs on T1 ... T5, performs a five-way merge from T1 ... T5 onto T6 until T5 is empty, then (without involving T6) a four-way merge onto T5, then a three-way merge onto T4, a two-way merge onto T3, and finally a copy operation from T1 onto T2. The next pass operates in the same way starting with a five-way merge to T1, and so on. Although this scheme seems to be inferior to Polyphase because at times it chooses to leave some sequences idle, and because it involves simple copy operations, it surprisingly is superior to Polyphase for (very) large files and for six or more sequences. Write a well structured program for the Cascade merge principle.

REFERENCES

2-1. B. K. Betz and Carter. *Proc. ACM National Conf. 14,* (1959), Paper 14.

2-2. R.W. Floyd. Treesort (Algorithms 113 and 243). *Comm. ACM, 5,* No. 8, (1962), 434, and *Comm. ACM, 7,* No. 12 (1964), 701.

2-3. R.L. Gilstad. Polyphase Merge Sorting - An Advanced Technique. *Proc. AFIPS Eastern Jt. Comp. Conf., 18,* (1960), 143-48.

2-4. C.A.R. Hoare. Proof of a Program: FIND. *Comm. ACM,13,* No. 1, (1970), 39-45.

2-5. -------- Proof of a Recursive Program: Quicksort.*Comp. J., 14,* No. 4 (1971), 391-95.

2-6. -------- Quicksort.*Comp.J., 5.* No.1 (1962), 10-15.

2-7. D.E. Knuth. *The Art of Computer Programming.* Vol. 3 (Reading, Mass.: Addison-Wesley, 1973).

2-8. --------*The Art of Computer Programming.* Vol 3, pp. 86-95.

2-9. --------*The Art of Computer Programming.* Vol 3, p. 289.

2-10. H. Lorin. A Guided Bibliography to Sorting. *IBM Syst.J., 10,* No. 3 (1971), 244-54.

2-11. D.L. Shell. A Highspeed Sorting Procedure. *Comm. ACM, 2,* No. 7 (1959), 30-32.

2-12. R.C. Singleton. An Efficient Algorithm for Sorting with Minimal Storage (Algorithm 347). *Comm. ACM, 12,* No. 3 (1969), 185.

2-13. M. H. Van Emden. Increasing the Efficiency of Quicksort (Algorithm 402). *Comm. ACM, 13,* No. 9 (1970), 563-66, 693.

2-14. J.W.J. Williams. Heapsort (Algorithm 232). *Comm. ACM, 7,* No. 6 (1964), 347-48.

3 RECURSIVE ALGORITHMS

3.1. INTRODUCTION

An object is said to be *recursive*, if it partially consists or is defined in terms of itself. Recursion is encountered not only in mathematics, but also in daily life. Who has never seen an advertising picture which contains itself?

Fig. 3.1 A recursive picture.

Recursion is a particularly powerful technique in mathematical definitions. A few familiar examples are those of natural numbers, tree structures, and of certain functions:

1. Natural numbers:
   (a) 0 is a natural number.
   (b) the successor of a natural number is a natural number.

2. Tree structures
   (a) O is a tree (called the empty tree).
   (b) If t1 and t2 are trees, then the structures consisting of a node with two descendant trees is also a tree.

3. The factorial function n! (for non-negative integers):
   (a) 0! = 1
   (b) n > 0: n! = n * (n - 1)!

The power of recursion evidently lies in the possibility of defining an infinite set of objects by a finite statement. In the same manner, an infinite number of computations can be described by a finite recursive program, even if this program contains no explicit repetitions. Recursive algorithms, however, are primarily appropriate when the problem to be solved, or the function to be computed, or the data structure to be processed are already defined in recursive terms. In general, a recursive program P can be expressed as a composition **P** of a set of statements **S** (not containing P) and P itself.

$$P \equiv \mathbf{P}[\mathbf{S}, P] \tag{3.1}$$

The necessary and sufficient tool for expressing programs recursively is the procedure or subroutine, for it allows a statement to be given a name by which this statement may be invoked. If a procedure P contains an explicit reference to itself, then it is said to be *directly recursive*; if P contains a reference to another procedure Q, which contains a (direct or indirect) reference to P, then P is said to be *indirectly recursive*. The use of recursion may therefore not be immediately apparent from the program text.

It is common to associate a set of local objects with a procedure, i.e., a set of variables, constants, types, and procedures which are defined locally to this procedure and have no existence or meaning outside this procedure. Each time such a procedure is activated recursively, a new set of local, bound variables is created. Although they have the same names as their corresponding elements in the set local to the previous instance of the procedure, their values are distinct, and any conflict in naming is avoided by the rules of scope of identifiers: the identifiers always refer to the most recently created set of variables. The same rule holds for procedure parameters, which by definition are bound to the procedure.

Like repetitive statements, recursive procedures introduce the possibility of non-terminating computations, and thereby also the necessity of considering the problem of termination. A fundamental requirement is evidently that the recursive calls of P are subjected to a condition B, which at some time becomes false. The scheme for recursive algorithms may therefore be expressed more precisely by either one of the following forms:

$$P \equiv \text{IF B THEN } \mathbf{P}[\mathbf{S}, P] \text{ END} \tag{3.2}$$

$$P \equiv \mathbf{P}[\mathbf{S}, \text{IF B THEN P END}] \tag{3.3}$$

For repetitions, the basic technique of demonstrating termination consists of

1. defining a function f(x) (x shall be the set of variables), such that $f(x) \leq 0$ implies the

terminating condition (of the while or repeat clause), and

2. proving that f(x) decreases during each repetition step.

In the same manner, termination of a recursion can be proved by showing that each execution of P decreases some f(x), and that $f(x) \leq 0$ implies ~B. A particularly evident way to ensure termination is to associate a (value) parameter, say n, with P, and to recursively call P with n-1 as parameter value. Substituting $n > 0$ for B then guarantees termination. This may be expressed by the following program schemata:

$$P(n) \equiv \text{IF } n > 0 \text{ THEN } \mathbf{P}[S, P(n-1)] \text{ END} \tag{3.4}$$

$$P(n) \equiv \mathbf{P}[S, \text{IF } n > 0 \text{ THEN } P(n-1) \text{ END}] \tag{3.5}$$

In practical applications it is mandatory to show that the ultimate depth of recursion is not only finite, but that it is actually quite *small*. The reason is that upon each recursive activation of a procedure P some amount of storage is required to accommodate its variables. In addition to these local variables, the current state of the computation must be recorded in order to be retrievable when the new activation of P is terminated and the old one has to be resumed. We have already encountered this situation in the development of the procedure *Quicksort* in Chap. 2. It was discovered that by naively composing the program out of a statement that splits the n items into two partitions and of two recursive calls sorting the two partitions, the depth of recursion may in the worst case approach n. By a clever reassessment of the situation, it was possible to limit the depth to log n. The difference between n and log n is sufficient to convert a case highly inappropriate for recursion into one in which recursion is perfectly practical.

3.2. WHEN NOT TO USE RECURSION

Recursive algorithms are particularly appropriate when the underlying problem or the data to be treated are defined in recursive terms. This does not mean, however, that such recursive definitions guarantee that a recursive algorithm is the best way to solve the problem. In fact, the explanation of the concept of recursive algorithm by such inappropriate examples has been a chief cause of creating widespread apprehension and antipathy toward the use of recursion in programming, and of equating recursion with inefficiency.

Programs in which the use of algorithmic recursion is to be avoided can be characterized by a schema which exhibits the pattern of their composition. The schema is that of (3.6) and, equivalently, of (3.7). Its characteristic is that there is only a single call of P either at the end (or the beginning) of the composition.

$$P \equiv \text{IF B THEN } \mathbf{S}\text{; P END} \tag{3.6}$$
$$P \equiv \mathbf{S}\text{; IF B THEN P END} \tag{3.7}$$

These schemata are natural in those cases in which values are to be computed that are defined in terms of simple recurrence relations. Let us look at the well-known example of the factorial numbers $f_i = i!$:

$$\begin{aligned} i &= 0, 1, 2, 3, 4, 5, \ldots \\ f_i &= 1, 1, 2, 6, 24, 120, \ldots \end{aligned} \tag{3.8}$$

The first number is explicitly defined as $f_0 = 1$, whereas the subsequent numbers are defined recursively in terms of their predecessor:

$$f_{i+1} = (i+1) * f_i \tag{3.9}$$

This recurrence relation suggests a recursive algorithm to compute the n th factorial number. If we introduce the two variables I and F to denote the values i and f_i at the i th level of recursion, we find the computation necessary to proceed to the next numbers in the sequences (3.8) to be

$$I := I + 1;\ F := I * F \tag{3.10}$$

and, substituting (3.10) for Σ in (3.6), we obtain the recursive program

```
P ≡ IF I < n THEN I := I + 1; F := I * F; P END        (3.11)
I := 0; F := 1; P
```

The first line of (3.11) is expressed in terms of our conventional programming notation as

```
PROCEDURE P;
BEGIN
  IF I < n THEN I := I + 1; F := I*F; P END              (3.12)
END P
```

A more frequently used, but essentially equivalent, form is the one given in (3.13). P is replaced by a so-called *function procedure*, i.e., a procedure with which a resulting value is explicitly associated, and which therefore may be used directly as a constituent of expressions. The variable F therefore becomes superfluous; and the role of I is taken over by the explicit procedure parameter.

```
PROCEDURE F(I: INTEGER): INTEGER;
BEGIN
  IF I > 0 THEN RETURN I * F(I - 1) ELSE RETURN 1 END          (3.13)
END F
```

It now is plain that in this example recursion can be replaced quite simply by iteration. This is expressed by the program

```
I := 0;  F := 1;
WHILE I < n DO I := I + 1; F := I*F END                         (3.14)
```

In general, programs corresponding to the schemata (3.6) or (3.7) should be transcribed into one according to schema (3.15)

```
P ≡ [x := x0; WHILE B DO S END]                                 (3.15)
```

There also exist more complicated recursive composition schemes that can and should be translated into an iterative form. An example is the computation of the *Fibonacci numbers* which are defined by the recurrence relation

$$fib_{n+1} = fib_n + fib_{n-1} \quad \text{for } n > 0 \qquad (3.16)$$

and $fib_1 = 1$, $fib_0 = 0$. A direct, naive transcription leads to the recursive program

```
PROCEDURE Fib(n: INTEGER): INTEGER;                             (3.17)
BEGIN
  IF n = 0 THEN RETURN 0
  ELSIF n = 1 THEN RETURN 1
  ELSE RETURN Fib(n-1) + Fib(n-2)
  END
END Fib
```

Computation of fib_n by a call Fib(n) causes this function procedure to be activated recursively. How often? We notice that each call with n > 1 leads to 2 further calls, i.e., the total number of calls grows exponentially (see Fig. 3.2). Such a program is clearly impractical.

But fortunately the Fibonacci numbers can be computed by an iterative scheme that avoids the recomputation of the same values by use of auxiliary variables such that $x = fib_i$ and $y = fib_{i-1}$.

```
i := 1; x := 1; y := 0;                                         (3.18)
WHILE i < n DO  z := x;  x := x + y;  y := z;  i := i + 1  END
```

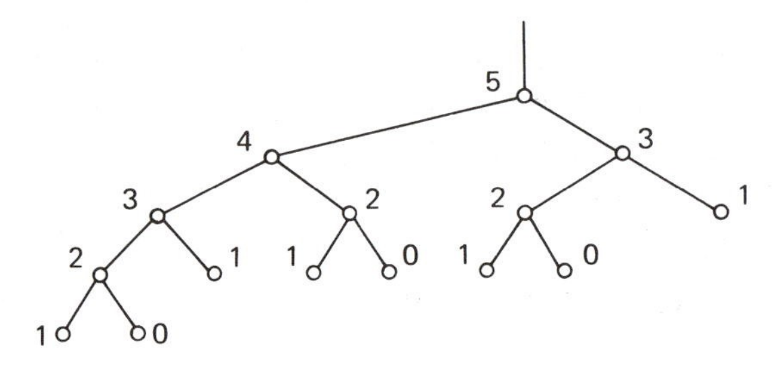

Fig. 3.2 The 15 calls of Fib(5).

Note: The assignments to x, y, z may be expressed by two assignments only without a need for the auxiliary variable z: x := x + y; y := x - y.

Thus, the lesson to be drawn is to avoid the use of recursion when there is an obvious solution by iteration. This, however, should not lead to shying away from recursion at any price. There are many good applications of recursion, as the following paragraphs and chapters will demonstrate. The fact that implementions of recursive procedures on essentially non-recursive machines exist proves that for practical purposes every recursive program can be transformed into a purely iterative one. This, however, involves the explicit handling of a recursion stack, and these operations will often obscure the essence of a program to such an extent that it becomes most difficult to comprehend. The lesson is that algorithms which by their nature are recursive rather than iterative should be formulated as recursive procedures. In order to appreciate this point, the reader is referred to Programs 2.10 and 2.11 for a comparison.

The remaining part of this chapter is devoted to the development of some recursive programs in situations in which recursion is justifiably appropriate. Also Chap. 4 makes extensive use of recursion in cases in which the underlying data structures let the choice of recursive solutions appear obvious and natural.

3.3. TWO EXAMPLES OF RECURSIVE PROGRAMS

The attractive graphic pattern shown in Fig. 3.5 consists of the superposition of five curves. These curves follow a regular pattern and suggest that they might be drawn by a display or a plotter under control of a computer. Our goal is to discover the recursion schema, according to which the drawing program might be constructed. Inspection reveals that three of the superimposed curves have the shapes shown in Fig. 3.3; we denote them by H_1, H_2, and H_3. The figures show that H_{i+1} is obtained by the composition of four instances of H_i of half size and appropriate rotation, and by tying together the four H_i by three connecting lines. Notice that H_1 may be considered as consisting of four instances of an empty H_0 connected by three straight lines. H_i is called the *Hilbert curve* of order i after its inventor, the mathematician D. Hilbert (1891).

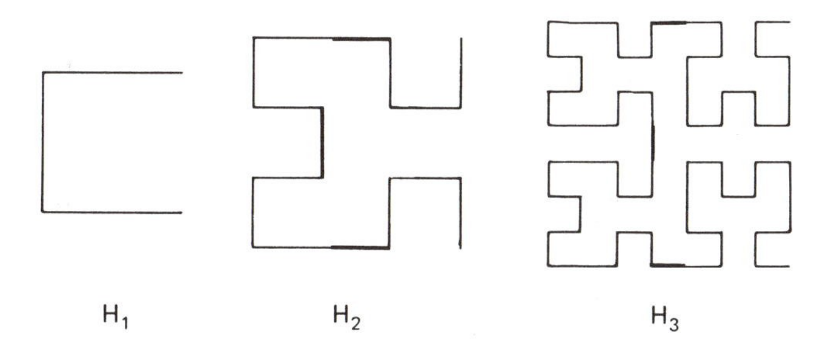

Fig. 3.3 Hilbert curves of order 1, 2, and 3.

Since each curve H_i consists of four half-sized copies of H_{i-1}, we express the procedure for drawing H_i as a composition of four calls for drawing H_{i-1} in half size and appropriate rotation. For the purpose of illustration we denote the four parts by A, B, C, and D, and the routines drawing the interconnecting lines by arrows pointing in the corresponding direction. Then the following recursion scheme emerges (see Fig. 3.3).

$$
\begin{aligned}
A:&\quad D \leftarrow\ A \downarrow\ A \rightarrow\ B \\
B:&\quad C \uparrow\ B \rightarrow\ B \downarrow\ A \\
C:&\quad B \rightarrow\ C \uparrow\ C \leftarrow\ D \\
D:&\quad A \downarrow\ D \leftarrow\ D \uparrow\ C
\end{aligned}
\tag{3.19}
$$

Let us assume that for drawing line segments we have at our disposal a procedure *line* which moves a drawing pen in a given direction by a given distance. For our convenience, we assume that the direction be indicated by an integer parameter i as 45i degrees. If the length of the unit line is denoted by u, the procedure corresponding to the scheme A is readily expressed by using recursive activations of analogously designed procedures B and D and of itself.

```
PROCEDURE A(i: INTEGER);
```

```
BEGIN
  IF i > 0 THEN
    D(i-1); line(4, u);
    A(i-1); line(6, u);
    A(i-1); line(0, u);                                    (3.20)
    B(i-1)
  END
END A
```

This procedure is initiated by the main program once for every Hilbert curve to be superimposed. The main program determines the initial point of the curve, i.e., the initial coordinates of the pen denoted by Px and Py, and the unit increment u. The square in which the curves are drawn is placed into the middle of the page with given width and height. These parameters as well as the drawing procedure line are taken from a module called *LineDrawing.* The entire program draws the n Hilbert curves H_i ... H_n (see Program 3.1 and Fig. 3.4).

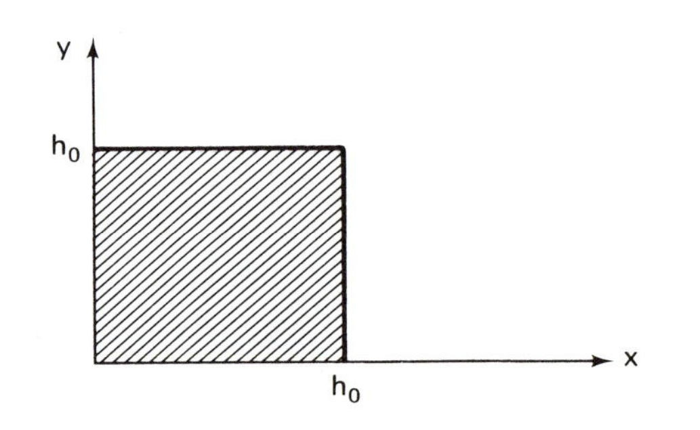

Fig. 3.4 The unit frame.

```
MODULE Hilbert;
 FROM Terminal IMPORT Read;
 FROM LineDrawing IMPORT width, height, Px, Py, clear, line;

 CONST SquareSize = 512;

 VAR i,x0,y0,u: CARDINAL; ch: CHAR;

 PROCEDURE A(i: CARDINAL);
 BEGIN
  IF i > 0 THEN
   D(i-1); line(4,u); A(i-1); line(6,u);
   A(i-1); line(0,u); B(i-1)
  END
 END A;

 PROCEDURE B(i: CARDINAL);
 BEGIN
  IF i > 0 THEN
```

```
      C(i-1); line(2,u); B(i-1); line(0,u);
      B(i-1); line(6,u); A(i-1)
    END
  END B;

  PROCEDURE C(i: CARDINAL);
  BEGIN
    IF i > 0 THEN
      B(i-1); line(0,u); C(i-1); line(2,u);
      C(i-1); line(4,u); D(i-1)
    END
  END C;

  PROCEDURE D(i: CARDINAL);
  BEGIN
    IF i > 0 THEN
      A(i-1); line(6,u); D(i-1); line(4,u);
      D(i-1); line(2,u); C(i-1)
    END
  END D;

BEGIN clear;
  x0 := width DIV 2; y0 := height DIV 2;
  u := SquareSize; i := 0;
  REPEAT i := i+1; u := u DIV 2;
    x0 := x0 + (u DIV 2); y0 := y0 + (u DIV 2);
    Px := x0; Py := y0; A(i); Read(ch)
  UNTIL (ch = 33C) OR (i = 6);
  clear
END Hilbert.
```

Program 3.1 Hilbert Curves.

A similar but slightly more complex and aesthetically more sophisticated example is shown in Fig. 3.7. This pattern is again obtained by superimposing several curves, two of which are shown in Fig. 3.6. S_i called the *Sierpinski curve* of order i. What is its recursion scheme? One is tempted to single out the leaf S_i as a basic building block, possibly with one edge left off. But this does not lead to a solution. The principal difference between Sierpinski curves and Hilbert curves is that Sierpinski curves are closed (without crossovers). This implies that the basic recursion scheme must be an open curve and that the four parts are connected by links not belonging to the recusion pattern itself. Indeed, these links consist of the four straight lines in the outermost four corners, drawn with thicker lines in Fig. 3.6. They may be regarded as belonging to a non-empty initial curve S_0, which is a square standing on one corner. Now the recursion schema is readily established. The four constituent patterns are again denoted by A, B, C, and D, and the connecting lines are drawn explicitly. Notice that the four recursion patterns are indeed identical except for 90 degree rotations.

Fig. 3.5 Hilbert curves $H_1 \ldots H_5$.

The base pattern of the Sierpinski curves is

$$S: \quad A \searrow B \swarrow C \nwarrow D \nearrow \tag{3.21}$$

and the recursion patterns are (horizontal and vertical arrows denote lines of double length.)

A: A ↘ B → D ↗ A
B: B ↙ C ↓ A ↘ B
C: C ↖ D ← B ↙ C (3.22)
D: D ↗ A ↑ C ↖ D

If we use the same primitives for drawing as in the Hilbert curve example, the above recursion scheme is transformed without difficulties into a (directly and indirectly) recursive algorithm.

```
PROCEDURE A(k: INTEGER);                                   (3.23)
BEGIN
 IF k > 0 THEN
  A(k-1); line(7, h); B(k-1); line(0, 2*h);
  D(k-1); line(1, h); A(k-1)
 END
END A
```

This procedure is derived from the first line of the recursion scheme (3.22). Procedures corresponding to the patterns B, C, and D are derived analogously. The main program is composed according to pattern (3.21). Its task is to set the initial values for the drawing coordinates and to determine the unit line length h according to the size of the plane, as shown in Program 3.2. The result of executing this program with n = 4 is shown in Fig. 3.7.

The elegance of the use of recursion in these exampes is obvious and convincing. The correctness of the programs can readily be deduced from their structure and composition patterns. Moreover, the use of an explicit level parameter according to schema (3.5) guarantees termination since the depth of recursion cannot become greater than n. In contrast to this recursive formulation, equivalent programs that avoid the explicit use of recursion are extremely cumbersome and obscure. Trying to understand the programs shown in [3-3] should easily convince the reader of this..

```
MODULE Sierpinski;
 FROM Terminal IMPORT Read;
 FROM LineDrawing IMPORT width, height, Px, Py, clear, line;

 CONST SquareSize = 512;

 VAR i,h,x0,y0: CARDINAL; ch: CHAR;

 PROCEDURE A(k: CARDINAL);
 BEGIN
  IF k > 0 THEN
   A(k-1); line(7, h); B(k-1); line(0, 2*h);
   D(k-1); line(1, h); A(k-1)
  END
 END A;

 PROCEDURE B(k: CARDINAL);
```

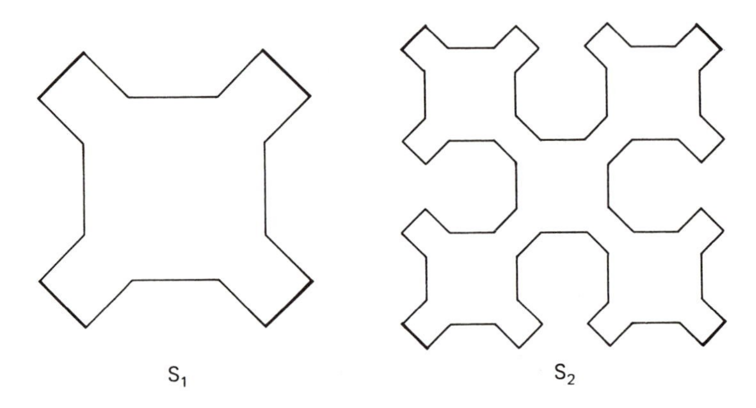

Fig. 3.6 Sierpinski curves of orders 1 and 2.

```
BEGIN
 IF k > 0 THEN
  B(k-1); line(5, h); C(k-1); line(6, 2*h);
  A(k-1); line(7, h); B(k-1)
 END
END B;

PROCEDURE C(k: CARDINAL);
BEGIN
 IF k > 0 THEN
  C(k-1); line(3, h); D(k-1); line(4, 2*h);
  B(k-1); line(5, h); C(k-1)
 END
END C;

PROCEDURE D(k: CARDINAL);
BEGIN
 IF k > 0 THEN
  D(k-1); line(1, h); A(k-1); line(2, 2*h);
  C(k-1); line(3, h); D(k-1)
 END
```

```
 END D;

BEGIN clear; i := 0; h := SquareSize DIV 4;;
  x0 := CARDINAL(width) DIV 2; y0 := CARDINAL(height) DIV 2 + h;
  REPEAT i := i+1; x0 := x0-h;
   h := h DIV 2; y0 := y0+h; Px := x0; Py := y0;
   A(i); line(7,h); B(i); line(5,h);
   C(i); line(3,h); D(i); line(1,h); Read(ch)
  UNTIL (i = 6) OR (ch = 33C);
  clear
END Sierpinski.
```

Program 3.2 Sierpinski Curves.

Fig. 3.7 Sierpinski curves $S_1 \ldots S_4$.

3.4. BACKTRACKING ALGORITHMS

A particularly intriguing programming endeavor is the subject of so-called *general problem solving.* The task is to determine algorithms for finding solutions to specific problems not by following a fixed rule of computation, but by trial and error. The common pattern is to decompose the *trial-and-error* process onto partial tasks. Often these tasks are most naturally expressed in recursive terms and consist of the exploration of a finite number of subtasks. We may generally view the entire process as a trial or search process that gradually builds up and scans (prunes) a tree of subtasks. In many problems this search tree grows very rapidly, often exponentially, depending on a given parameter. The search effort increases accordingly. Frequently, the search tree can be pruned by the use of heuristics only, thereby reducing computation to tolerable bounds.

It is not our aim to discuss general heuristic rules in this text. Rather, the general principle of breaking up such problem-solving tasks into subtasks and the application of recursion is to be the subject of this chapter. We start out by demonstrating the underlying technique by using an example, namely, the well known *knight's tour.*

Given is a n×n board with n^2 fields. A knight -- being allowed to move according to the rules of chess -- is placed on the field with initial coordinates x_0, y_0. The problem is to find a covering of the entire board, if there exists one, i.e. to compute a tour of n^2-1 moves such that every field of the board is visited exactly once.

The obvious way to reduce the problem of covering n^2 fields is to consider the problem of either performing a next move or finding out that none is possible. Let us therefore define an algorithm trying to perform a next move. A first approach is shown in (3.24).

```
PROCEDURE TryNextMove;                                                  (3.24)
BEGIN initialize selection of moves;
  REPEAT select next candidate from list of next moves;
    IF acceptable THEN
      record move;
      IF board not full THEN
        TryNextMove;
        IF not successful THEN erase previous recording END
      END
    END
  UNTIL (move was successful) OR (no more candidates)
END TryNextMove
```

If we wish to be more precise in describing this algorithm, we are forced to make some decisions on data representation. An obvious step is to represent the board by a matrix, say h. Let us also introduce a type to denote index values:

```
TYPE index = [1 .. n];                                                  (3.25)
VAR h: ARRAY index, index OF INTEGER
```

The decision to represent each field of the board by an integer instead of a Boolean value denoting occupation is because we wish to keep track of the history of successive board occupations. The following convention is an obvious choice:

$$h[x,y] = 0: \text{ field } \langle x,y \rangle \text{ has not been visited}$$
$$h[x,y] = i: \text{ field } \langle x,y \rangle \text{ has been visited in the i th move } (1 \le i \le n^2) \quad (3.26)$$

The next decision concerns the choice of appropriate parameters. They are to determine the starting conditions for the next move and also to report on its success. The former task is adequately solved by specifying the coordinates x,y from which the move is to be made, and by specifying the number i of the move (for recording purposes). The latter task requires a Boolean result parameter with the meaning *the move was successful.*

Which statements can now be refined on the basis of these decisions? Certainly *board not full* can be expressed as $i < n^2$. Moreover, if we introduce two local variables u and v to stand for the coordinates of possible move destinations determined according to the jump pattern of knights, then the predicate *acceptable* can be expressed as the logical conjunction of the conditions that the new field lies on the board, i.e. $1 \le u \le n$ and $1 \le v \le n$, and that it had not been visited previously, i.e., $h_{uv} = 0$.

The operation of recording the legal move is expressed by the assignment $h_{uv} := i$, and the cancellation of this recording as $h_{uv} := 0$. If a local variable q1 is introduced and used as the result parameter in the recursive call of this algorithm, then q1 may be substituted for *move was successful.* Thereby we arrive at the formulation shown in (3.27).

```
PROCEDURE Try(i: INTEGER; x, y: index; VAR q: BOOLEAN);
  VAR u, v: INTEGER; q1: BOOLEAN;
BEGIN initialize selection of moves;
  REPEAT let <u,v> be the coordinates of the next move
    as defined by the rules of chess;
    IF (1 <= u) & (u <= n) & (1 <= v) & (v <= n) & (h[u,v] = 0) THEN
      h[u,v] := i;
      IF i < n*n THEN Try(i+1, u, v, q1);
        IF ~q1 THEN h[u,v] := 0 ELSE q1 := TRUE END
      END
    END
  UNTIL q1 OR no more candidates;
  q := q1
END Try
```

Just one more refinement step will lead us to a program expressed fully in terms of our basic programming notation. We should note that so far the program was developed completely independently of the laws governing the jumps of the knight. This delaying of considerations of particularities of the problem was quite deliberate. But now is the time to take them into account.

Given a starting coordinate pair x,y, there are eight potential candidates u,v of the destination. They are numbered 1 to 8 in Fig. 3.8. A simple method of obtaining u,v from x,y is by addition of the coordinate differences stored in either an array of difference pairs or in two arrays of single differences. Let these arrays be denoted by *dx* and *dy,* appropriately initialized. Then an index k may be used to number the *next candidate.* The details are shown in Program 3.3. The recursive procedure is initiated by a call with the coordinates x0, y0 of that field as parameters from which the tour is to start. This field must be given the value 1; all others are to be marked free.

h[x0, y0] := 1; try(2, x0, y0, q)

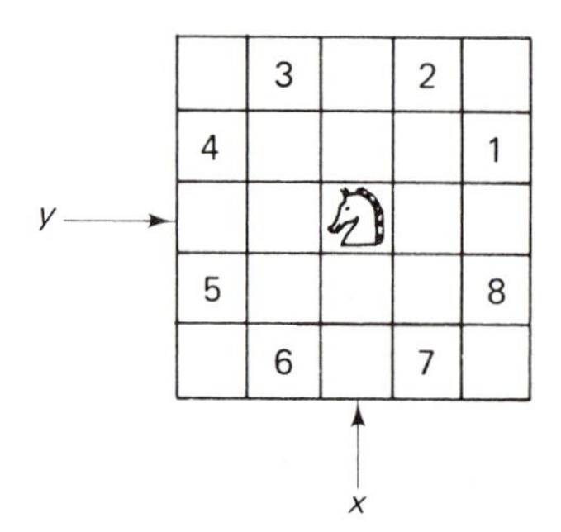

Fig. 3.8 The eight possible moves of a knight.

One further detail must not be overlooked: A variable h_{uv} does exist only if both u and v lie within the index range 1 ... n. Consequently, the expression in (3.27), substituted for *acceptable* in (3.24), is valid only if its first four constituent terms are true. It is therefore relevant that the term $h_{uv} = 0$ appears last. Table 3.1 indicates solutions obtained with initial positions <3,3>, <2,4> for n = 5 and <1,1> for n = 6.

```
MODULE KnightsTour;
  FROM InOut IMPORT
    ReadInt, Done, WriteInt, WriteString, WriteLn;

  VAR i, j, n, Nsqr: INTEGER; q: BOOLEAN;
    dx, dy: ARRAY [1 .. 8] OF INTEGER;
    h: ARRAY [1 .. 8], [1 .. 8] OF INTEGER;

  PROCEDURE Try(i, x, y: INTEGER; VAR q: BOOLEAN);
    VAR k, u, v: INTEGER; q1: BOOLEAN;
  BEGIN k := 0;
    REPEAT k := k+1; q1 := FALSE;
      u := x + dx[k]; v := y + dy[k];
      IF (1 <= u) & (u <= n) & (1 <= v) & (v <= n) & (h[u,v] = 0) THEN
        h[u,v] := i;
        IF i < Nsqr THEN Try(i+1, u, v, q1);
          IF ~q1 THEN h[u,v] := 0 END
        ELSE q1 := TRUE
```

```
        END
      END
    UNTIL q1 OR (k = 8);
    q := q1
  END Try;

BEGIN
  a[1] := 2; a[2] := 1; a[3] := -1; a[4] := -2;
  a[5] := -2; a[6] := -1; a[7] := 1; a[8] := 2;
  b[1] := 1; b[2] := 2; b[3] := 2; b[4] := 1;
  b[5] := -1; b[6] := -2; b[7] := -2; b[8] := -1;
  LOOP ReadInt(n);
    IF ~Done THEN EXIT END ;
    FOR i := 1 TO n DO
      FOR j := 1 TO n DO h[i,j] := 0 END
    END ;
    ReadInt(i); ReadInt(j); WriteLn;
    Nsqr := n*n; h[1,1] := 1; Try(2, i, j, q);
    IF q THEN
      FOR i := 1 TO n DO
        FOR j := 1 TO n DO WriteInt(h[i,j], 5) END ;
        WriteLn
      END
    ELSE WriteString(" no path"); WriteLn
    END
  END
END KnightsTour.
```

Program 3.3 Knight's Tour.

23	10	15	4	25
16	5	24	9	14
11	22	1	18	3
6	17	20	13	8
21	12	7	2	19

23	4	9	14	25
10	15	24	1	8
5	22	3	18	13
16	11	20	7	2
21	6	17	12	19

1	16	7	26	11	14
34	25	12	15	6	27
17	2	33	8	13	10
32	35	24	21	28	5
23	18	3	30	9	20
36	31	22	19	4	29

Table 3.1 Three Knights' Tours.

What abstractions can now be made from this example? Which pattern does it exhibit that is typical for this kind of problem-solving algorithms? What does it teach us? The characteristic feature is that steps toward the total solution are attempted and recorded that may later be taken back and erased in the recordings when it is discovered that the step does not possibly lead to the total solution, that the step leads into a dead-end street. This action is called *backtracking.* The general pattern (3.28) is derived from (3.24), assuming that the number of potential candidates in each step is finite.

```
PROCEDURE Try;
BEGIN intialize selection of candidates;
  REPEAT select next;                                                    (3.28)
    IF acceptable THEN
      record it;
      IF solution incomplete THEN Try;
        IF not successful THEN cancel recording END
      END
    END
  UNTIL successful OR no more candidates
END Try
```

Actual programs may, of course, assume various derivative forms of schema (3.28). A frequently encountered pattern uses an explicit level parameter indicating the depth of recursion and allowing for a simple termination condition. If, moreover, at each step the number of candidates to be investigated is fixed, say m, then the derived schema (3.29) applies; it is to be invoked by the statement *Try(1).*

```
PROCEDURE Try(i: INTEGER);
   VAR k: INTEGER;
BEGIN k := 0;
  REPEAT k := k+1; select k-th candidate;
    IF acceptable THEN                                                   (3.29)
      record it;
      IF i < n THEN Try(i+1);
        IF not successful THEN cancel recording END
      END
    END
  UNTIL successful OR (k = m)
END Try
```

The remainder of this chapter is devoted to the treatment of three more examples. They display various incarnations of the abstract schema (3.29) and are included as further illustrations of the appropriate use of recursion.

3.5. THE EIGHT QUEENS PROBLEM

The problem of the eight queens is a well-known example of the use of trial-and-error methods and of backtracking algorithms. It was investiated by C.F. Gauss in 1850, but he did not completely solve it. This should not surprise anyone. After all, the characteristic property of these problems is that they defy analytic solution. Instead, they require large amounts of exacting labor, patience, and accuracy. Such algorithms have therefore gained relevance almost exclusively through the automatic computer, which possesses these properties to a much higher degree than people, and even geniuses, do.

The eight queens poblem is stated as follows (see also [3-4]): Eight queens are to be placed on a chess board in such a way that no queen checks against any other queen. Using schema (3.29) as a template, we readily obtain the following crude version of a solution:

```
PROCEDURE Try(i: INTEGER);
BEGIN
  initialize selection of positions for i-th queen:
  REPEAT make next selection;
    IF safe THEN SetQueen;
      IF i < 8 THEN Try(i+1);
        IF not successful THEN RemoveQueen END
      END
    END
  UNTIL successful OR no more positions
END Try
```

In order to proceed, it is necessary to make some commitments concerning the data representation. Since we know from the rules of chess that a queen checks all other figures lying in either the same column, row, or diagonal on the board, we infer that each column may contain one and only one queen, and that the choice of a position for the i th queen may be restricted to the i th column. The parameter i therefore becomes the column index, and the selection process for positions then ranges over the eight possible values for a row index j.

There remains the question of representing the eight queens on the board. An obvious choice would again be a square matrix to represent the board, but a little inspection reveals that such a representation would lead to fairly cumbersome operations for checking the availability of positions. This is highly undesirable since it is the most frequently executed operation. We should therefore choose a data representation which makes checking as simple as possible. The best recipe is to represent as directly as possible that information which is truly relevant and most often used. In our case this is not the position of the queens, but whether or not a queen has already been placed along each row and diagonals. (We already know that exactly one is placed in each column k for $1 \leq k \leq i$). This leads to the following choice of variables:

```
VAR x: ARRAY [1 .. 8] OF INTEGER;
```

```
a: ARRAY [1 .. 8] OF BOOLEAN;
b: ARRAY [b1 .. b2] OF BOOLEAN;                                    (3.31)
c: ARRAY [c1 .. c2] OF BOOLEAN;
```

where

x_i denotes the position of the queen in the i th column;

a_j means "no queen lies in the j th row";

b_k means "no queen occupies the k th /-diagonal;

c_k means "no queen sits on the k th \-diagonal.

The choice for index bounds b1, b2, c1, c2 is dictated by the way that indices of b and c are computed; we note that in a /-diagonal all fields have the same sums of their coordinates i and j, and that in a \-diagonal the coordinate differences i-j are constant. The appropriate solution is shown in Program 3.4. Given these data, the statement *SetQueen* is elaborated to

```
x[i] := j; a[j] := FALSE; b[i+j] := FALSE; c[i-j] := FALSE          (3.32)
```

the statement *RemoveQueen* is refined into

```
a[j] := TRUE; b[i+j] := TRUE; c[i-j] := TRUE                         (3.33)
```

and the condition *safe* is fulfilled if the field <i,j> lies in in a row and in diagonals which are still free. Hence, it can be expressed by the logical expression

```
a[j] & b[i+j] & c[i-j]                                              (3.34)
```

This completes the development of this algorithm, that is shown in full as Program 3.4. The computed solution is x = (1, 5, 8, 6, 3, 7, 2, 4) and is shown in Fig. 3.9.

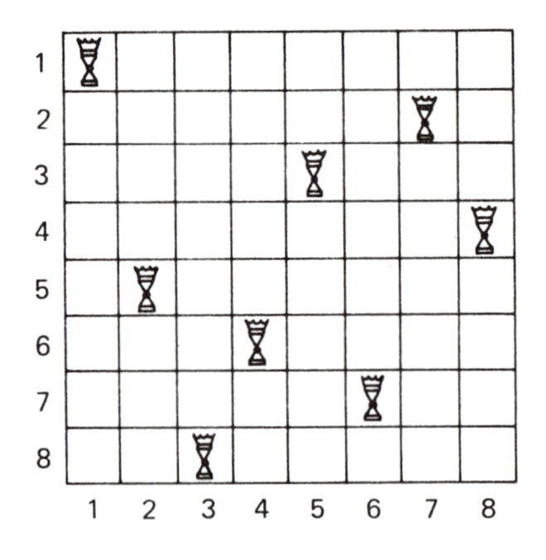

Fig. 3.9 A solution to the eight queens problem.

```
MODULE Queens;
  FROM InOut IMPORT WriteInt, WriteLn;
```

```
VAR i: INTEGER; q: BOOLEAN;
  a: ARRAY [ 1 .. 8] OF BOOLEAN;
  b: ARRAY [ 2 .. 16] OF BOOLEAN;
  c: ARRAY [-7 .. 7] OF BOOLEAN;
  x: ARRAY [ 1 .. 8] OF INTEGER;

 PROCEDURE Try(i: INTEGER; VAR q: BOOLEAN);
   VAR j: INTEGER;
 BEGIN j := 0;
   REPEAT j := j+1; q := FALSE;
    IF a[j] & b[i+j] & c[i-j] THEN
      x[i] := j;
      a[j] := FALSE; b[i+j] := FALSE; c[i-j] := FALSE;
      IF i < 8 THEN
        Try(i+1, q);
        IF ~q THEN
          a[j] := TRUE; b[i+j] := TRUE; c[i-j] := TRUE
        END
      ELSE q := TRUE
      END
    END
   UNTIL q OR (j = 8)
 END Try;

BEGIN
  FOR i :=  1 TO  8 DO a[i] := TRUE END ;
  FOR i :=  2 TO 16 DO b[i] := TRUE END ;
  FOR i := -7 TO  7 DO c[i] := TRUE END ;
  Try(1,q);
  FOR i := 1 TO 8 DO WriteInt(x[i], 4) END ;
  WriteLn
END Queens.
```

Program 3.4 Eight Queens.

Before we abandon the context of the chess board, the eight queens example is to serve as an illustration of an important extension of the trial-and-error algorithm. The extension is -- in general terms -- to find not only one, but *all* solutions to a posed problem.

The extension is easily accommodated. We are to recall the fact that the generation of candidates must progress in a systematic manner that guarantees no candidate is generated more than once. This property of the algorithm corresponds to a search of the candidate tree in a systematic fashion in which every node is visited exactly once. It allows -- once a solution is found and duly recorded -- merely to proceed to the next candidate delivered by the systematic selection process. The general schema is derived from (3.29) and shown in (3.35)

```
PROCEDURE Try(i: INTEGER);
  VAR k: INTEGER;
BEGIN                                                        (3.35)
  FOR k := 1 TO m DO
    select k th candidate;
    IF acceptable THEN record it;
      IF i < n THEN Try(i+1) ELSE print solution END ;
      cancel recording
    END
  END
END Try
```

Note that because of the simplification of the termination condition of the selection process to the single term k = m, the repeat statement is appropriately replaced by a for statement. It comes as a surprise that the search for all possible solutions is realized by a simpler program than the search for a single solution.

The extended algorithm to determine all 92 solutions of the eight queens problem is shown in Program 3.5. Actually, there are only 12 significantly differing solutions; our program does not recognize symmetries. The 12 solutions generated first are listed in Table 3.2. The numbers n to the right indicate the frequency of execution of the test for safe fields. Its average over all 92 solutions is 161.

```
MODULE AllQueens;
  FROM InOut IMPORT WriteInt, WriteLn;

VAR i: INTEGER;
  a: ARRAY [1 .. 8] OF BOOLEAN;
  b: ARRAY [2 .. 16] OF BOOLEAN;
  c: ARRAY [-7 .. 7] OF BOOLEAN;
  x: ARRAY [1 .. 8] OF INTEGER;

 PROCEDURE print;
   VAR k: INTEGER;
 BEGIN
   FOR k := 1 TO 8 DO WriteInt(x[k], 4) END ;
   WriteLn
 END print;

 PROCEDURE Try(i: INTEGER);
   VAR j: INTEGER;
 BEGIN
   FOR j := 1 TO 8 DO
     IF a[j] & b[i+j] & c[i-j] THEN
       x[i] := j;
       a[j] := FALSE; b[i+j] := FALSE; c[i-j] := FALSE;
       IF i < 8 THEN Try(i+1) ELSE print END ;
```

```
      a[j] := TRUE; b[i+j] := TRUE; c[i-j] := TRUE
    END
  END
END Try;

BEGIN
  FOR i :=  1 TO  8 DO a[i] := TRUE END ;
  FOR i :=  2 TO 16 DO b[i] := TRUE END ;
  FOR i := -7 TO  7 DO c[i] := TRUE END ;
  Try(1)
END AllQueens.
```

Program 3.5 Eight Queens.

x_1	x_2	x_3	x_4	x_5	x_6	x_7	x_8	n
1	5	8	6	3	7	2	4	876
1	6	8	3	7	4	2	5	264
1	7	4	6	8	2	5	3	200
1	7	5	8	2	4	6	3	136
2	4	6	8	3	1	7	5	504
2	5	7	1	3	8	6	4	400
2	5	7	4	1	8	6	3	072
2	6	1	7	4	8	3	5	280
2	6	8	3	1	4	7	5	240
2	7	3	6	8	5	1	4	264
2	7	5	8	1	4	6	3	160
2	8	6	1	3	5	7	4	336

Table 3.2 Twelve Solutions to the Eight Queens Problem.

3.6. THE STABLE MARRIAGE PROBLEM

Assume that two disjoint sets A and B of equal size n are given. Find a set of n pairs <a,b> such that a in A and b in B satisfy some constrains. Many different criteria for such pairs exist; one of them is the rule called the *stable marriage rule.*

Assume that A is a set of men and B is a set of women. Each man and each women has stated distinct preferences for their possible partners. If the n couples are chosen such that there exists a man and a woman who are not married, but who would both prefer each other to their actual marriage partners, then the assignment is unstable. If no such pair exists, it is called *stable.* This situation characterizes many related problems in which assignments have to be made according to preferences such as, for example, the choice of a school by students, the choice of recruits by different branches of the armed services, etc. The example of marriages is particularly intuitive; note, however, that the stated list of preferences is invariant and does not change after a particular assignment has been made. This assumption simplifies the problem, but it also represents a grave distortion of reality (called abstraction).

One way to search for a solution is to try pairing off members of the two sets one after the other until the two sets are exhausted. Setting out to find all stable assignments, we can readily sketch a solution by using the program schema (3.35) as a template. Let *Try(m)* denote the algorithm to find a partner for man m, and let this search proceed in the order of the man's list of stated preferences. The first version based on these assumptions is (3.36).

```
PROCEDURE Try(m: man);
  VAR r: rank;
BEGIN                                                        (3.36)
  FOR r := 1 TO n DO
    pick the r th preference of man m;
    IF acceptable THEN record the marriage;
      IF m is not last man THEN Try(successor(m))
      ELSE record the stable set
      END;
      cancel the marriage
    END
  END
END Try
```

Again, we have arrived at the point where we cannot proceed without first making some decisions about data representation. We introduce three scalar types, and, for reasons of simplicity, let their values be the integers 1 to n. Although the three types are formally identical, their designation by distinct names enhances clarity significantly. In particular, it can more easily be made evident what a variable stands for.

```
TYPE man   = [1 .. n];
     woman = [1 .. n];                                       (3.37)
     rank  = [1 .. n]
```

The initial data are represented by two matrices that indicate the men's and women's preferences.

VAR wmr: ARRAY man, rank OF woman
mwr: ARRAY woman, rank OF man (3.38)

Accordingly, wmr_m denotes the preference list of man m, i.e., $wmr_{m,r}$ is the woman who occupies the r th rank in the list of man m. Similarly, mwr_w is the preference list of woman w, and $mwr_{w,r}$ is her r th choice. A sample data set is shown in Table 3.3.

r =	1	2	3	4	5	6	7	8			1	2	3	4	5	6	7	8
m = 1	7	2	6	5	1	3	8	4	w =	1	4	6	2	5	8	1	3	7
2	4	3	2	6	8	1	7	5		2	8	5	3	1	6	7	4	2
3	3	2	4	1	8	5	7	6		3	6	8	1	2	3	4	7	5
4	3	8	4	2	5	6	7	1		4	3	2	4	7	6	8	5	1
5	8	3	4	5	6	1	7	2		5	6	3	1	4	5	7	2	8
6	8	7	5	2	4	3	1	6		6	2	1	3	8	7	4	6	5
7	2	4	6	3	1	7	5	8		7	3	5	7	2	4	1	8	6
8	6	1	4	2	7	5	3	8		8	7	2	8	4	5	6	3	1

Table 3.3 Sample Input Data for *wmr* and *mwr*

The result is represented by an array of women x, such that x_m denotes the partner of man m. In order to maintain symmetry between men and women, an additional array y is introduced, such that y_w denotes the partner of woman w.

VAR x: ARRAY man OF woman;
y: ARRAY woman OF man (3.39)

Actually, y is redundant, since it represents information that is already present through the existence of x. In fact, the relations

$$x_{y_w} = w, \quad y_{x_m} = m \tag{3.40}$$

hold for all m and w who are married. Thus, the value y_w could be determined by a simple search of x; the array y, however, clearly improves the efficiency of the algorithm. The information represented by x and y is needed to determine stability of a proposed set of marriages. Since this set is constructed stepwise by marrying individuals and testing stability after each proposed marriage, x and y are needed even before all their components are defined. In order to keep track of defined components, we may introduce Boolean arrays

singlem: ARRAY man OF BOOLEAN
singlew: ARRAY woman OF BOOLEAN (3.41)

with the meaning that $singlem_m$ implies that x_m is defined and $singlew_w$ implies that y_w is defined. An inspection of the proposed algorithm, however, quickly reveals that the marital status of a man is determined by the value m through the relation

~singlem[k] = k < m (3.42)

This suggests that the array *singlem* be omitted; accordingly, we will simplify the name *singlew* to *single*. These conventions lead to the refinement shown in (3.43). The predicate *acceptable* can be refined into the conjunction of *single* and *stable*, where *stable* is a function to be still further elaborated.

```
PROCEDURE Try(m: man);
  VAR r: rank; w: woman;
BEGIN                                                              (3.43)
  FOR r := 1 TO n DO
    w := wmr[m,r];
    IF single[w] & stable THEN
      x[m] := w; y[w] := m; single[w] := FALSE;
      IF m < n THEN Try(successor(m)) ELSE record set END ;
      single[w] := TRUE
    END
  END
END Try
```

At this point, the strong similarity of this solution with Program 3.5 is still noticeable. The crucial task is now the refinement of the algorithm to determine stability. Unfortunately, it is not possible to represent stability by such a simple expression as the safety of a queen's position in Program 3.5 The first detail that should be kept in mind is that stability follows by definition from comparisons of ranks. The ranks of men or women, however, are nowhere explicitly available in our collection of data established so far. Surely, the rank of woman w in the mind of man m can be computed, but only by a costly search of w in wmr_m. Since the computation of stability is a very frequent operation, it is advisable to make this information more directly accessible. To this end, we introduce the two matrices

```
rmw: ARRAY man, woman OF rank;                                     (3.44)
rwm: ARRAY woman, man OF rank
```

such that $rmw_{m,w}$ denotes woman w's rank in the preference list of man m, and $rwm_{w,m}$ denotes the rank of man m in the list of w. It is plain that the values of these auxiliary arrays are constant and can initially be determined from the values of wmr and mwr.

The process of determining the predicate *stable* now proceeds strictly according to its original definition. Recall that we are trying the feasibility of marrying m and w, where w = $wmr_{m,r}$, i.e., w is man m's r th choice. Being optimistic, we first presume that stability still prevails, and then we set out to find possible sources of trouble. Where could they be hidden? There are two symmetrical possibilities:

1. There might be a women *pw*, preferred to w by m, who herself prefers m over her husband.
2. There might be a man *pm*, preferred to m by w, who himself prefers w over his wife.

Pursuing trouble source 1, we compare ranks $rwm_{pw,m}$ and $rwm_{pw,y_{pw}}$ for all women preferred to w by m, i.e. for all pw = $wmr_{m,i}$ such that $i < r$. We happen to know that all these candidate women are already married because, were anyone of them still single, m would have picked her beforehand. The described process can be formulated by a simple linear search; s denotes stability.

```
s := TRUE; i := 1;
WHILE (i < r) & s DO
  pw := wmr[m,i]; i := i+1;                                        (3.45)
  IF ~single[pw] THEN s := rwm[pw,m] > rwm[pw, y[pw]] END
END
```

Hunting for trouble source 2, we must investigate all candidates pm who are preferred by w to their current assignation m, i.e., all preferred men $pm = mwr_{w,i}$ such that $i < rwm_{w,m}$. In analogy to tracing trouble source 1, comparison between ranks $rmw_{pm,w}$ and $rmw_{pm,x_{pm}}$ is necessary. We must be careful, however, to omit comparisons involving x_{pm} where pm is still single. The necessary safeguard is a test *pm < m*, since we know that all men preceding m are already married.

The complete algorithm is shown in Program 3.6. Table 3.4 specifies the nine computed stable solutions from input data wmr and mwr given in Table 3.3.

```
MODULE Marriage;
 FROM InOut IMPORT
  ReadCard, WriteCard, WriteLn;

 CONST n = 8;
 TYPE man = [1 .. n];
   woman = [1 .. n];
   rank = [1 .. n];

 VAR m: man; w: woman; r: rank;
   wmr: ARRAY man, rank OF woman;
   mwr: ARRAY woman, rank OF man;
   rmw: ARRAY man, woman OF rank;
   rwm: ARRAY woman, man OF rank;
   x: ARRAY man OF woman;
   y: ARRAY woman OF man;
   single: ARRAY woman OF BOOLEAN;
   h: CARDINAL;

 PROCEDURE print;
  VAR m: man; rm, rw: CARDINAL;
 BEGIN rm := 0; rw := 0;
  FOR m := 1 TO n DO
   WriteCard(x[m], 4);
   rm := rmw[m, x[m]] + rm; rw := rwm[x[m], m] + rw
  END ;
  WriteCard(rm, 8); WriteCard(rw, 4); WriteLn
 END print;

 PROCEDURE stable(m: man; w: woman; r: rank): BOOLEAN;
```

```
    VAR pm: man; pw: woman;
      i, lim: rank; S: BOOLEAN;
  BEGIN S := TRUE; i := 1;
    WHILE (i < r) & S DO
      pw := wmr[m,i]; i := i+1;
      IF ~single[pw] THEN S := rwm[pw,m] > rwm[pw, y[pw]] END
    END ;
    i := 1; lim := rwm[w,m];
    WHILE (i < lim) & S DO
      pm := mwr[w,i]; i := i+1;
      IF pm < m THEN S := rmw[pm,w] > rmw[pm, x[pm]] END
    END ;
    RETURN S
  END stable;

  PROCEDURE Try(m: man);
    VAR w: woman; r: rank;
  BEGIN
    FOR r := 1 TO n DO w := wmr[m,r];
      IF single[w] & stable(m,w,r) THEN
        x[m] := w; y[w] := m; single[w] := FALSE;
        IF m < n THEN Try(m+1) ELSE print END ;
        single[w] := TRUE
      END
    END
  END Try;

BEGIN
  FOR m := 1 TO n DO
    FOR r := 1 TO n DO
      ReadCard(h); wmr[m,r] := h; rmw[m, wmr[m,r]] := r
    END
  END ;
  FOR w := 1 TO n DO
    single[w] := TRUE;
    FOR r := 1 TO n DO
      ReadCard(h); mwr[w,r] := h; rwm[w, mwr[w,r]] := r
    END
  END ;
  Try(1)
END Marriage.
```

Program 3.6 Stable Marriages.

This algorithm is based on a straightforward backtracking scheme. Its efficiency primarily depends on the sophistication of the solution tree pruning scheme. A somewhat faster, but more complex and less transparent algorithm has been presented by McVitie and

Wilson [3-1 and 3-2], who also have extended it to the case of sets (of men and women) of unequal size.

Algorithms of the kind of the last two examples, which generate all possible solutions to a problem (given certain constraints), are often used to select one or several of the solutions that are optimal in some sense. In the present example, one might, for instance, be interested in the solution that on the average best satisfies the men, or the women, or everyone.

Notice that Table 3.4 indicates the sums of the ranks of all women in the preference lists of their husbands, and the sums of the ranks of all men in the preference lists of their wives. These are the values

$$\begin{aligned} rm &= \mathbf{S}\, m: 1 \le m \le n: rmw_{m,x_m} \\ rw &= \mathbf{S}\, m: 1 \le m \le n: rwm_{x_m,m} \end{aligned} \quad (3.46)$$

	x_1	x_2	x_3	x_4	x_5	x_6	x_7	x_8	rm	rw	c
1	7	4	3	8	1	5	2	6	16	32	21
2	2	4	3	8	1	5	7	6	22	27	449
3	2	4	3	1	7	5	8	6	31	20	59
4	6	4	3	8	1	5	7	2	26	22	62
5	6	4	3	1	7	5	8	2	35	15	47
6	6	3	4	8	1	5	7	2	29	20	143
7	6	3	4	1	7	5	8	2	38	13	47
8	3	6	4	8	1	5	7	2	34	18	758
9	3	6	4	1	7	5	8	2	43	11	34

c = number of evaluations of stability.
Solution 1 = male optimal solution; solution 9 = female optimal solution.

Table 3.4 Result of the Stable Marriage Problem.

The solution with the least value rm is called the *male-optimal* stable solution; the one with the smallest rw is the *female-optimal* stable solution. It lies in the nature of the chosen search strategy that good solutions from the men's point of view are generated first and the good solutions from the women's perspective appear toward the end. In this sense, the algorithm is based toward the male population. This can quickly be changed by systematically interchanging the role of men and women, i.e., by merely interchanging mwr with *wmr* and interchanging rmw with *rwm*.

We refrain from extending this program further and leave the incorporation of a search for an optimal solution to the next and last example of a backtracking algorithm.

3.7. THE OPTIMAL SELECTION PROBLEM

The last example of a backtracking algorithm is a logical extension of the previous two examples represented by the general schema (3.35). First we were using the principle of backtracking to find a *single* solution to a given problem. This was exemplified by the knight's tour and the eight queens. Then we tackled the goal of finding *all* solutions to a given problem; the examples were those of the eight queens and the stable marriages. Now we wish to find *an optimal* solution.

To this end, it is necessary to generate all possible solutions, and in the course of generating them to retain the one that is optimal in some specific sense. Assuming that optimality is defined in terms of some positive valued function f(s), the algorithm is derived from schema (3.35) by replacing the statement *print solution* by the statement

```
IF f(solution) > f(optimum) THEN optimum := solution END        (3.47)
```

The variable *optimum* records the best solution so far encountered. Naturally, it has to be properly initialized; morever, it is customary to record to value *f(optimum)* by another variable in order to avoid its frequent recomputation.

An example of the general problem of finding an optimal solution to a given problem follows: We choose the important and frequently encountered problem of finding an optimal selection out of a given set of objects subject to constraints. Selections that constitute acceptable solutions are gradually built up by investigating individual objects from the base set. A procedure *Try* describes the process of investigating the suitability of one individual object, and it is called recursively (to investigate the next object) until all objects have been considered.

We note that the consideration of each object (called candidates in previous examples) has two possible outcomes, namely, either the inclusion of the investigated object in the current selection or its exclusion. This makes the use of a repeat or for statement inappropriate; instead, the two cases may as well be explicitly written out. This is shown in (3.48), assuming that the objects are numbered 1, 2, ... , n.

```
PROCEDURE Try(i: INTEGER);
BEGIN
  IF inclusion is acceptable THEN include i th object;
    IF i < n THEN Try(i+1) ELSE check optimality END ;
    eliminate i th object                    (3.48)
  END ;
  IF exclusion is acceptable THEN
    IF i < n THEN Try(i+1) ELSE check optimality END
  END
END Try
```

From this pattern it is evident that there are 2^n possible sets; clearly, appropriate acceptability criteria must be employed to reduce the number of investigated candidates very drastically. In order to elucidate this process, let us choose a concrete example for a selection problem: Let each of the n objects $a_1, \dots, a_n$ be characterized by its weight and its value. Let the optimal set be the one with the largest sum of the values of its components, and let the constraint be a limit on the sum of their weight. This is a problem well known to all travellers who pack suitcases by selecting from n items in such a way that their total value is optimal and that their total weight does not exceed a specific allowance.

We are now in a position to decide upon the representation of the given facts in terms of data. The choices of (3.49) are easily derived from the foregoing developments.

```
TYPE index = [1 .. n];
    object = RECORD weight, value: INTEGER END ;

VAR obj: ARRAY index OF object;                                  (3.49)
   limw, totv, maxv: INTEGER;
   s, opts: SET OF index
```

The variables *limw* and *totv* denote the weight limit and the total value of all n objects. These two values are actually constant during the entire selection process. s represents the current selection of objects in which each object is represented by its name (index). *opts* is the optimal selection so far encountered, and *maxv* is its value.

Which are now the criteria for acceptability of an object for the current selection? If we consider *inclusion*, then an object is selectable, if it fits into the weight allowance. If it does not fit, we may stop trying to add further objects to the current selection. If, however, we consider *exclusion*, then the criterion for acceptability, i.e., for the continuation of building up the current selection, is that the total value which is still achievable after this exclusion is not less than the value of the optimum so far encountered. For, if it is less, continuation of the search, although it may produce some solution, will not yield the optimal solution. Hence any further search on the current path is fruitless. From these two conditions we determine the relevant quantities to be computed for each step in the selection process:

1. The total weight *tw* of the selection s so far made.
2. The still achievable value *av* of the current selection s.

These two entities are appropriately represented as parameters of the procedure *Try*. The condition *inclusion is acceptable* in (3.48) can now be formulated as

```
tw + a[i].weight ≤ limw                                          (3.50)
```

and the subsequent check for optimality as

```
IF av > maxv THEN (*new optimum, record it*)
  opts := s; maxv := av                                          (3.51)
END
```

The last assignment is based on the reasoning that the achievable value is the achieved value, once all n objects have been dealt with. The condition *exclusion is acceptable* in (3.48) is expressed by

```
av - a[i].value > maxv                                           (3.52)
```

Since it is used again thereafter, the value av - a[i].value is given the name *av1* in order to circumvent its reevaluation.

The entire program now follows from (3.48) through (3.52) with the addition of appropriate initialization statements for the global variables. The ease of expressing inclusion and exclusion from the set s by use of set operators is noteworthy. The results of execution of Program 3.7 with weight allowances ranging from 10 to 120 are listed in Table 3.5.

```
MODULE Selection;
 (*find optimal selection of objects under constraint*)
 FROM InOut IMPORT
  ReadCard, Write, WriteCard, WriteString, WriteLn;

 CONST n = 10;
 TYPE index = [1 .. n];
    object = RECORD value, weight: CARDINAL END ;
    ObjSet = SET OF index;

 VAR i: index;
  obj: ARRAY index OF object;
  limw, totv, maxv: CARDINAL;
  s, opts: ObjSet;
  WeightInc, WeightLimit: CARDINAL;
  tick: ARRAY [FALSE .. TRUE] OF CHAR;

 PROCEDURE Try(i: index; tw, av: CARDINAL);
  VAR av1: CARDINAL;
 BEGIN (*try inclusion*)
  IF tw + obj[i].weight <= limw THEN
   s := s + ObjSet{i};
   IF i < n THEN Try(i+1, tw + obj[i].weight, av)
   ELSIF av > maxv THEN maxv := av; opts := s
   END ;
   s := s - ObjSet{i}
  END ;
  (*try exclusion*)
  IF av > maxv + obj[i].value THEN
   IF i < n THEN Try(i+1, tw, av - obj[i].value)
   ELSE maxv := av - obj[i].value; opts := s
   END
  END
 END Try;

BEGIN totv := 0; limw := 0;
 tick[FALSE] := " "; tick[TRUE] := "*";
 FOR i := 1 TO n DO
```

```
      ReadCard(obj[i].weight); ReadCard(obj[i].value);
      totv := totv + obj[i].value
    END ;
    ReadCard(WeightInc); ReadCard(WeightLimit);
    WriteString("Weight");
    FOR i := 1 TO n DO WriteCard(obj[i].weight, 5) END ;
    WriteLn; WriteString("Value ");
    FOR i := 1 TO n DO WriteCard(obj[i].value, 5) END ;
    WriteLn;
    REPEAT limw := limw + WeightInc; maxv := 0;
      s := ObjSet{}; opts := ObjSet{}; Try(1, 0, totv);
      WriteCard(limw, 6);
      FOR i := 1 TO n DO
        WriteString("    "); Write(tick[i IN opts])
      END ;
      WriteCard(maxv, 8); WriteLn
    UNTIL limw >= WeightLimit
  END Selection.
```

PROGRAM 3.7 Optimal Selection.

Weight	10	11	12	13	14	15	16	17	18	19	
Value	18	20	17	19	25	21	27	23	25	24	
10	*										18
20							*				27
30					*		*				52
40	*				*		*				70
50	*	*		*			*				84
60	*	*	*	*	*						99
70	*	*			*		*		*		115
80	*	*	*		*		*	*			130
90	*	*			*		*		*	*	139
100	*	*		*	*		*	*	*		157
110	*	*	*	*	*	*	*		*		172
120	*	*			*	*	*	*	*	*	183

Table 3.5 Sample Output from Optimal Selection Program.

This backtracking scheme with a limitation factor curtailing the growth of the potential search tree is also known as *branch and bound* algorithm.

EXERCISES

3.1 (Towers of Hanoi). Given are three rods and n disks of different sizes. The disks can be stacked up on the rods, thereby forming towers. Let the n disks initially be placed on rod A in the order of decreasing size, as shown in Fig. 3.10 for $n = 3$. The task is to move the n disks from rod A to rod C such that they are ordered in the original way. This has to be achieved under the constraints that

1. In each step exactly one disk is moved from one rod to another rod.
2. A disk may never be placed on top of a smaller disk.
3. Rod B may be used as an auxiliary store.

Find an algorithm that performs this task. Note that a tower may conveniently be considered as consisting of the single disk at the top, and the tower consisting of the remaining disks. Describe the algorithm as a recursive program.

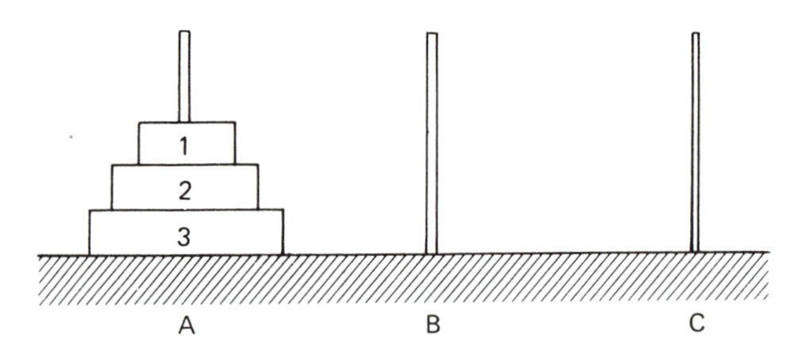

Fig. 3.10 Towers of Hanoi.

3.2. Write a procedure that generates all n! permutations of n elements $a_1, \dots, a_n$ *in situ,* i.e., without the aid of another array. Upon generating the next permutation, a parametric procedure Q is to be called which may, for instance, output the generated permutation.

Hint: Consider the task of generating all permutations of the elements $a_1, \dots, a_m$ as consisting of the m subtasks of generating all permutations of $a_1, \dots, a_{m-1}$ followed by a_m, where in the i th subtask the two elements a_i and a_m had initially been interchanged.

3.3. Deduce the recursion scheme of Fig. 3.11 which is a superposition of the four curves W_1, W_2, W_3, W_4. The structure is similar to that of the Sierpinski curves (3.21) and (3.22). From the recursion pattern, derive a recursive program that draws these curves.

3.4. Only 12 of the 92 solutions computed by the Eight Queens Program 3.5 are essentially different. The other ones can be derived by reflections about axes or the center point. Devise a program that determines the 12 principal solutions. Note that, for example,

Fig. 3.11 W-curves of order 1 through 4.

the search in column 1 may be restricted to positions 1-4.

3.5 Change the Stable Marriage Program so that it determines the optimal solution (male or female). It therefore becomes a branch and bound program of the type represented by Program 3.7.

3.6 A certain railway company serves n stations $S_1, \dots, S_n$. It intends to improve its customer information service by computerized information terminals. A customer types in his departure station S_A and his destination S_D, and he is supposed to be (immediately) given the schedule of the train connections with minimum total time of the journey. Devise a program to compute the desired information. Assume that the timetable (which is your data bank) is provided in a suitable data structure containing departure (= arrival) times of all available trains. Naturally, not all stations are connected by direct lines (see also Exercise 1.6).

3.7 The Ackermann Function A is defined for all non-negative integer arguments m and n as follows:

$$
\begin{aligned}
A(0, n) &= n + 1 \\
A(m, 0) &= A(m-1, 1) && (m > 0) \\
A(m, n) &= A(m-1, A(m, n-1)) && (m, n > 0)
\end{aligned}
$$

Design a program that computes A(m,n) without the use of recursion. As a guideline, use Program 2.11, the non-recusive version of Quicksort. Devise a set of rules for the transformation of recursive into iterative programs in general.

REFERENCES

3-1. D.G. McVitie and L.B. Wilson. The Stable Marriage Problem. *Comm. ACM, 14,* No. 7 (1971), 486-92.

3-2. -------. Stable Marriage Assignment for Unequal Sets.*Bit, 10,* (1970), 295-309.

3-3. Space Filling Curves, or How to Waste Time on a Plotter. *Software - Practice and Experience, 1,* No. 4 (1971), 403-40.

3-4. N. Wirth. Program Development by Stepwise Refinement. *Comm. ACM, 14,* No. 4 (1971), 221-27.

4 DYNAMIC INFORMATION STRUCTURES

4.1. RECURSIVE DATA TYPES

In Chap. 2 the array, record, and set structures were introduced as fundamental data structures. They are called fundamental because they constitute the building blocks out of which more complex structures are formed, and because in practice they do occur most frequently. The purpose of defining a data type, and of thereafter specifying that certain variables be of that type, is that the range of values assumed by these variables, and therefore their storage pattern, is fixed once and for all. Hence, variables declared in this way are said to be *static*. However, there are many problems which involve far more complicated information structures. The characteristic of these problems is that not only the values but also the structures of variables change during the computation. They are therefore called *dynamic* structures. Naturally, the components of such structures are -- at some level of resolution -- static, i.e., of one of the fundamental data types. This chapter is devoted to the construction, analysis, and management of dynamic information structures.

It is noteworthy that there exist some close analogies between the methods used for structuring algorithms and those for structuring data. As with all analogies, there remain some differences, but a comparison of structuring methods for programs and data is nevertheless illuminating.

The elementary, unstructured statement is the *assignment* of an expression's value to a variable. Its corresponding member in the family of data structures is the scalar, unstructured type. These two are the atomic building blocks for composite statements and data types. The simplest structures, obtained through enumeration or sequencing, are the compound statement and the record structure. They both consist of a finite (usually small) number of explicitly enumerated components, which may themselves all be different from each other. If all components are identical, they need not be written out individually: we use the for statement and the array structure to indicate replication by a known, finite factor. A choice among two or more elements is expressed by the conditional or the case statement and by the variant record structure, respectively. And finally, a repetiton by an initially unknown (and potentially infinite) factor is expressed by the while or repeat statements. The corresponding data structure is the sequence (file), the simplest kind which allows the construction of types of infinite cardinality.

The question arises whether or not there exists a data structure that corresponds in a similar way to the procedure statement. Naturally, the most interesting and novel property of procedures in this respect is *recursion*. Values of such a recursive data type would contain one or more components belonging to the same type as itself, in analogy to a procedure containing one or more calls to itself. Like procedures, data type definitions might be directly or indirectly recursive.

A simple example of an object that would most appropriately be represented as a recursively defined type is the arithmetic expression found in programming languages. Recursion is used to reflect the possibility of nesting, i.e., of using parenthesized subexpressions as operands in expressions. Hence, let an expression here be defined informally as follows:

> An *expression* consists of a term, followed by an operator, followed by a term. (The two terms constitute the operands of the operator.) A *term* is either a variable -- represented by an identifier -- or an expression enclosed in parentheses.

A data type whose values represent such expressions can easily be described by using the tools already available with the addition of recursion:

```
TYPE expression = RECORD op: operator;
                    opd1, opd2: term
                  END

TYPE term =       RECORD                                    (4.1)
                    CASE t: BOOLEAN OF
                      TRUE: id: alfa |
                      FALSE: subex: expression
                    END
                  END
```

Note: Using Modula-2, we must use a case construction here, because the language does not permit an if construction within record definitions.

Hence, every variable of type *term* consists of two components, namely, the tagfield *t* and, if t is true, the field *id*, or of the field *subex* otherwise. Consider now, for example, the following four expressions:

```
1. x + y
2. x - (y * z)
3. (x + y) * (z - w)                                        (4.2)
4. (x/(y + z)) * w
```

These expressions may be visualized by the patterns in Fig. 4.1, which exhibit their nested, recursive structure, and they determine the layout or mapping of these expressions onto a store.

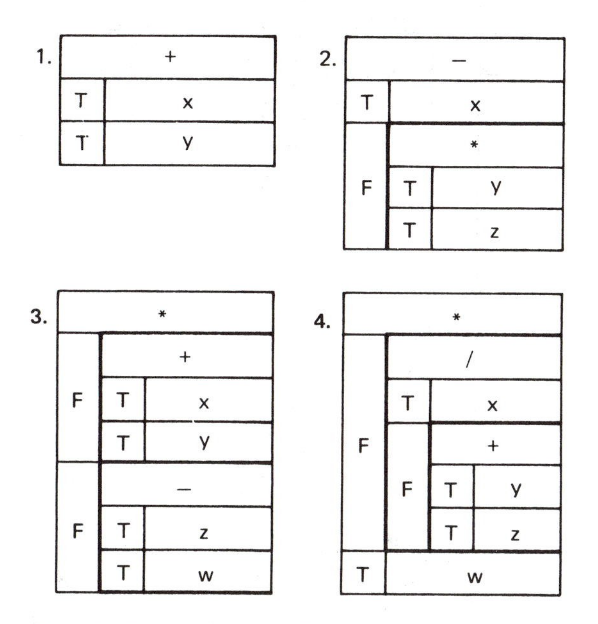

Fig. 4.1 Storage pattern for recursive record structures.

A second example of a recursive information structure is the family pedigree: Let a pedigree be defined by (the name of) a person and the two pedigrees of the parents. This definition leads inevitably to an infinite structure. Real pedigrees are bounded because at some level of ancestry information is missing. This can be taken into account by again using a variant structure as shown in (4.3).

```
TYPE ped =  RECORD
              CASE known: BOOLEAN OF
                TRUE: name: alfa; father, mother: ped |        (4.3)
                FALSE: (*empty*)
              END
            END
```

Note that every variable of type *ped* has at least one component, namely, the tagfield called *known*. If its value is TRUE, then there are three more fields; otherwise there is none. A particular value is shown here in the forms of a nested expression and of a diagram that may suggest a possible storage pattern (see Fig. 4.2).

(T, Ted, (T, Fred, (T, Adam, (F), (F)), (F)), (T, Mary, (F), (T, Eva, (F), (F)))

The important role of the variant facility becomes clear; it is the only means by which a recursive data structure can be bounded, and it is therefore an inevitable companion of every recursive definition. The analogy between program and data structuring concepts is particularly pronounced in this case. A conditional (or selective) statement must necessarily

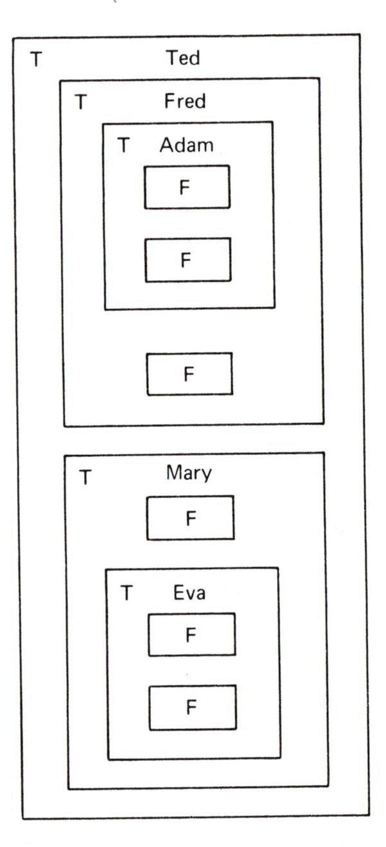

Fig. 4.2 An example of a recursive data structure.

be part of every recursive procedure in order that execution of the procedure can terminate. Termination of execution evidently corresponds to finite cardinality.

4.2. POINTERS

The characteristic property of recursive structures which clearly distinguishes them from the fundamental structures (arrays, records, sets) is their ability to vary in size. Hence, it is impossible to assign a fixed amount of storage to a recursively defined structure, and as a consequence a compiler cannot associate specific addresses to the components of such variables. The technique most commonly used to master this problem involves *dynamic allocation* of storage, i.e., allocation of store to individual components at the time when they come into existence during program execution, instead of at translation time. The compiler then allocates a fixed amount of storage to hold the address of the dynamically allocated component instead of the component itself. For instance, the pedigree illustrated in Fig. 4.2 would be represented by individual -- quite possibly noncontiguous -- records, one for each person. These persons are then linked by their addresses assigned to the respective *father* and *mother* fields. Graphically, this situation is best expressed by the use of arrows or pointers (Fig. 4.3).

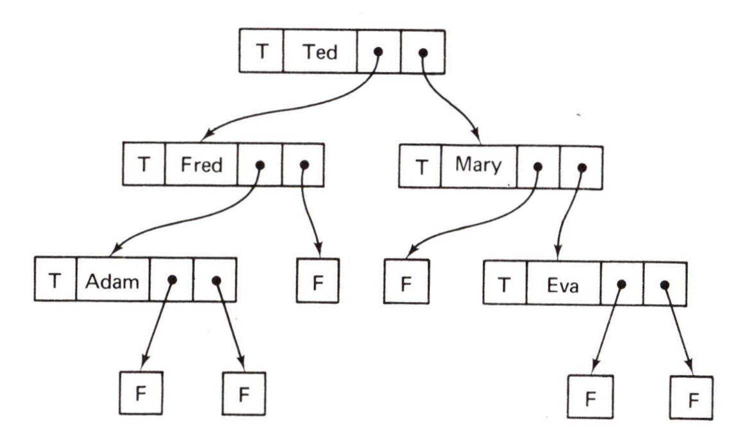

Fig. 4.3 Structure linked by pointers.

It must be emphasized that the use of pointers to implement recursive structures is merely a technique. The programmer need not be aware of their existence. Storage may be allocated automatically the first time a new component is referenced. However, if the technique of using references or pointers is made explicit, more general data structures can be constructed than those definable by purely recursive data definiton. In particular, it is then possible to define potentially infinite or circular structures and to dictate that certain structures are shared. It has therefore become common in advanced programming languages to make possible the explicit manipulation of references to data in additon to the data themeselves. This implies that a clear notational distinction must exist between data and references to data and that consequently data types must be introduced whose values are *pointers* (references) to other data. The notation we use for this purpose is the following:

```
TYPE T = POINTER TO T0                                                (4.4)
```

The type declaration (4.4) expresses that values of type T are pointers to data of type T0. It is fundamentally important that the type of elements pointed to is evident from the declaration of T. We say that T is *bound* to T0. This binding distinguishes pointers in higher-level languages from addresses in assembly codes, and it is a most important facility to increase security in programming through redundancy of the underlying notation.

Values of pointer types are generated whenever a data item is dynamically allocated. We will adhere to the convention that such an occasion be explicitly mentioned at all times. This is in contrast to the situation in which the first time that an item is mentioned it is automatically allocated. For this purpose, we introduce a procedure *Allocate.* Given a pointer variable p of type T, the statement *Allocate(p)* effectively allocates a variable of type T0 and assigns the pointer referencing this new variable to p (see Fig. 4.4). The pointer value itself can now be referred to as p (i.e., as the value of the pointer variable p). In contrast, the variable which is referenced by p is denoted by p↑.

Note: If Allocate is a procedure imported from a general storage management module, it is necessary to specify the size of the variable explicitly by a second parameter:

```
Allocate(p, SIZE(T0))                                                 (4.5)
```

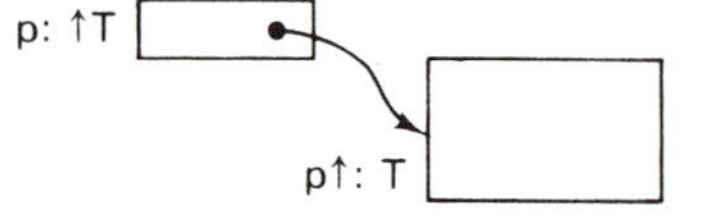

Fig. 4.4 Dynamic allocation of variable.

It was mentioned above that a variant component is essential in every recursive type to ensure finite cardinality. The example of the family predigree is of a pattern that exhibits a most frequently occurring constellation [see (4.3)], namely, the case in which the tag field is two-valued (Boolean) and in which its value being false implies the absence of any further components. This is expressed by the declaration schema (4.6).

```
TYPE T =   RECORD                                                     (4.6)
             CASE terminal: BOOLEAN OF
               TRUE:  S(T) |
               FALSE: (*empty*)
             END
           END
```

S(T) denotes a sequence of field definitions which includes one or more fields of type T, thereby ensuring recursivity. All structures of a type patterned after (4.6) exhibit a tree (or list) structure similar to that shown in Fig. 4.3. Its peculiar property is that it contains pointers to data components with a tag field only, i.e., without further relevant information. The implementation technique using pointers suggests an easy way of saving storage space by letting the tag information be included in the pointer value itself. The common solution is to extend the range of values of all pointer types by a single value that is pointing to no element at all. We denote this value by the special symbol NIL, and we understand that NIL is automatically an element of all pointer types declared. This extension of the range of pointer values explains why finite structures may be generated without the explicit presence of variants (conditions) in their (recursive) declaration.

The new formulations of the data types declared in (4.1) and (4.3) -- based on explicit pointers -- are given in (4.7) and (4.8), respectively. Note that in the latter case (which originally corresponded to schema (4.6)) the variant record component has vanished, since *~p.known* is now expressed as *p = NIL*. The renaming of the type *ped* to *person* reflects the difference in the viewpoint brought about by the introduction of explicit pointer values. Instead of first considering the given structure in its entirety and then investigating its substructure and its components, attention is focused on the components in the first place, and their interrelationship (represented by pointers) is not evident from any fixed declaration.

```
TYPE termPtr =     POINTER TO term;                    (4.7)

TYPE expression =  RECORD op: operator;
                     opdl, opd2: termPtr
                   END ;

TYPE term =        RECORD
                     CASE t: BOOLEAN OF
                       TRUE:  id: alfa |
                       FALSE: sub: expression
                     END
                   END

TYPE PersonPtr =   POINTER TO Person                   (4.8)

TYPE person =      RECORD name: alfa;
                     father, mother:  PersonPtr
                   END
```

The data structure representing the pedigree shown in Figs. 4.2 and 4.3 is again shown in Fig. 4.5 in which pointers to unknown persons are denoted by NIL. The resulting improvement in storage economy is obvious.

Again referring to Fig. 4.5, assume that Fred and Mary are siblings, i.e., have the same father and mother. This situation is easily expressed by replacing the two NIL values in the respective fields of the two records. An implementation that hides the concept of pointers or uses a different technique of storage handling would force the programmer to represent the ancestor records of Adam and Eve twice. Although in accessing their data for inspection it does not matter whether the two fathers (and the two mothers) are duplicated or represented by a single record, the difference is essential when selective updating is permitted. Treating pointers as explicit data items instead of as hidden implementation aids

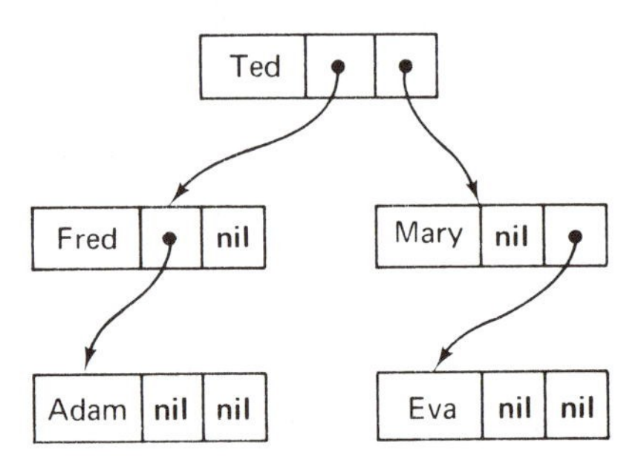

Fig. 4.5 Structure with NIL pointers.

allows the programmer to express clearly where storage sharing is intended and where it is not.

A further consequence of the explicitness of pointers is that it is possible to define and manipulate cyclic data structures. This additional flexibility yields, of course, not only increased power but also requires increased care by the programmer, because the manipulation of cyclic data structures may easily lead to nonterminating processes.

This phenomenon of power and flexibility being intimately coupled with the danger of misuse is well known in programming, and it particularly recalls the GOTO statement. Indeed, if the analogy between program structures and data structures is to be extended, the purely recursive data structure could well be placed at the level corresponding with the procedure, whereas the introduction of pointers is comparable to the use of GOTO statements. For, as the GOTO statement allows the construction of any kind of program pattern (including loops), so do pointers allow for the composition of any kind of data structure (including rings). The parallel development of corresponding program and data structures is shown in condensed form in Table 4.1.

Construction Pattern	Program Statement	Data Type
Atomic element	Assignment	Scalar type
Enumeration	Compound statement	Record type
Repetition (known factor)	For statement	Array type
Choice	Conditional statement	Type union (Variant record)
Repetition	While or repeat statement	Sequence type
Recursion	Procedure statement	Recursive data type
General graph	GO TO statement	Structure linked by pointers

Table 4.1 Correspondences of Program and Data Structures.

In Chap. 3, we have seen that iteration is a special case of recursion, and that a call of a recursive procedure P defined according to schema (4.9)

```
PROCEDURE P;
BEGIN
  IF B THEN P0; P END                                    (4.9)
```

```
END
```

where P0 is a statement not involving P, is equivalent to and replaceable by the iterative statement

```
WHILE B DO P0 END
```

The analogies outlined in Table 4.1 reveal that a similar relationship holds between recursive data types and the sequence. In fact, a recursive type defined according to the schema

```
TYPE T = RECORD
           CASE B: BOOLEAN OF                                    (4.10)
             TRUE: t0: T0; t: T |
             FALSE:
           END
         END
```

where T0 is a type not involving T, is equivalent and replaceable by a sequential data type

```
SEQUENCE OF T0
```

The remainder of this chapter is devoted to the generation and manipulation of data structures whose components are linked by explicit pointers. Structures with specific simple patterns are emphasized in particular; recipes for handling more complex structures may be derived from those for manipulating basic formations. These are the linear list or chained sequence -- the simplest case -- and trees. Our preoccupation with these building blocks of data structuring does not imply that more involved structures do not occur in practice. In fact, the following story appeared in a Zürich newspaper in July 1922 and is a proof that irregularity may even occur in cases which usually serve as examples for regular structures, such as (family) trees. The story tells of a man who laments the misery of his life in the following words:

I married a widow who had a grown-up daughter. My father, who visited us quite often, fell in love with my step-daughter and married her. Hence, my father became my son-in-law, and my step-daughter became my mother. Some months later, my wife gave birth to a son, who became the brother-in-law of my father as well as my uncle. The wife of my father, that is my stepdaughter, also had a son. Thereby, I got a brother and at the same time a grandson. My wife is my grandmother, since she is my mother's mother. Hence, I am my wife's husband and at the same time her step-grandson; in other words, I am my own grandfather.

4.3. LINEAR LISTS

4.3.1. Basic Operations

The simplest way to interrelate or link a set of elements is to line them up in a single list or queue. For, in this case, only a single link is needed for each element to refer to its successor.

Assume that a type *Node* and a type *Ptr* are defined as shown in (4.11). Every variable of this type consists of three components, namely, an identifying key, the pointer to its successor, and possibly further associated information omitted in (4.11).

```
TYPE Ptr =  POINTER TO Node;
TYPE Node = RECORD key: INTEGER;
                next: Ptr;                                    (4.11)
                data: ...
            END ;
VAR p, q: Ptr
```

A list of nodes, with a pointer to its first component being assigned to a variable p, is illustrated in Fig. 4.6. Probably the simplest operation to be performed with a list as shown in Fig. 4.6 is the insertion of an element at its head. First, an element of type *Node* is allocated, its reference (pointer) being assigned to an auxiliary pointer variable, say q. Thereafter, a simple reassignment of pointers completes the operation, which is programmed in (4.12). Note that the order of these three statements is essential.

```
Allocate(q, SIZE(Node)); q↑.next := p; p := q                 (4.12)
```

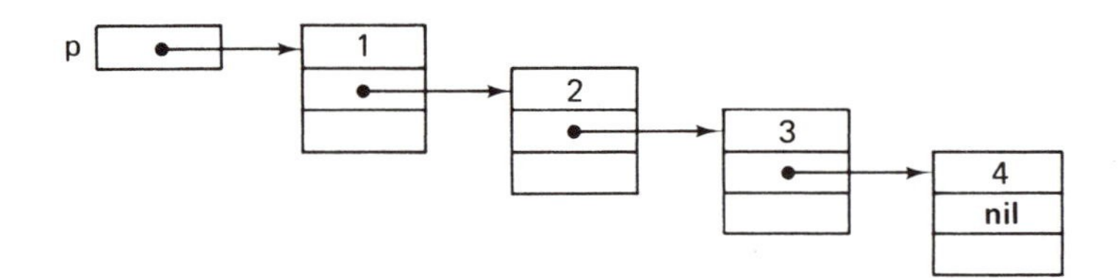

Fig. 4.6 Example of a list.

The operation of inserting an element at the head of a list immediately suggests how such a list can be generated: starting with the empty list, a heading element is added repeatedly. The process of *list generation* is expressed in (4.13); here the number of elements to be linked is n.

```
p := NIL; (*start with empty list*)
WHILE n > 0 DO
  Allocate(q, SIZE(Node)); q↑.next := p; p := q;              (4.13)
```

```
    q↑.key := n; n := n-1
END
```

This is the simplest way of forming a list. However, the resulting order of elements is the inverse of the order of their insertion. In some applications this is undesirable, and consequently, new elements must be appended at the end instead of the head of the list. Although the end can easily be determined by a scan of the list, this naive approach involves an effort that may as well be saved by using a second pointer, say q, always designating the last element. This method is, for example, applied in Program 4.4, which generates cross-references to a given text. Its disadvantage is that the first element inserted has to be treated differently from all later ones.

The explicit availability of pointers makes certain operations very simple which are otherwise cumbersome; among the elementary list operations are those of inserting and deleting elements (selective updating of a list), and, of course, the traversal of a list. We first investigate *list insertion.*

Assume that an element designated by a pointer (variable) q is to be inserted in a list *after* the element designated by the pointer p. The necessary pointer assignments are expressed in (4.14), and their effect is visualized by Fig. 4.7.

```
q↑.next := p↑.next; p↑.next := q                              (4.14)
```

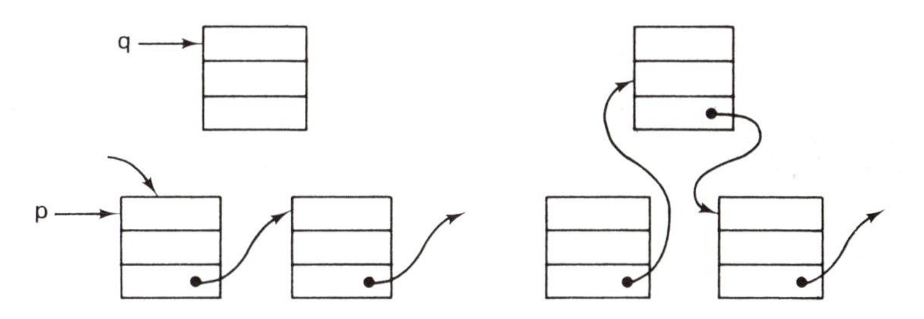

Fig. 4.7 List insertion after p↑.

If insertion *before* instead of after the designated element p↑ is desired, the unidirectional link chain seems to cause a problem, because it does not provide any kind of path to an element's predecessors. However, a simple trick solves our dilemma: it is expressed in (4.15) and illustrated in Fig. 4.8. Assume that the key of the new element is 8.

```
Allocate(q, SIZE(Node)); q↑ := p↑;                             (4.15)
p↑.key := k; p↑.next := q
```

The trick evidently consists of actually inserting a new component *after* p↑ and thereafter interchanging the values of the new element and p↑.

Next, we consider the process of *list deletion.* Deleting the successor of a p↑ is straightforward. In (4.16) it is shown in combination with the reinsertion of the deleted element at the head of another list (designated by q). Figure 4.9 illustrates the effect of statements (4.16) and shows that it is a cyclic exchange of three pointers.

$$r := p\uparrow.next;\ \ p\uparrow.next := r\uparrow.next;\ \ r\uparrow.next := q;\ \ q := r \tag{4.16}$$

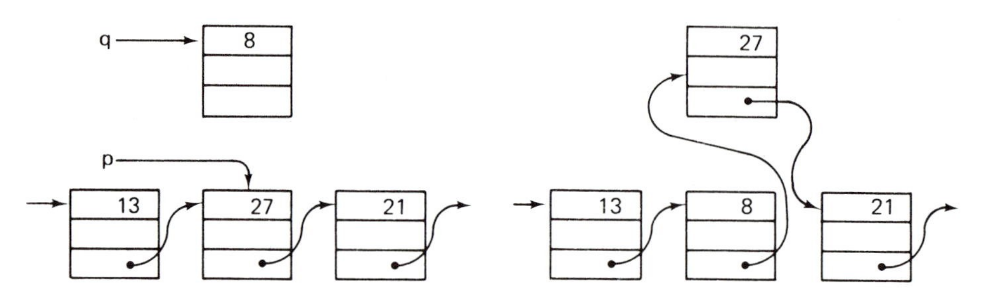

Fig. 4.8 List insertion before p↑.

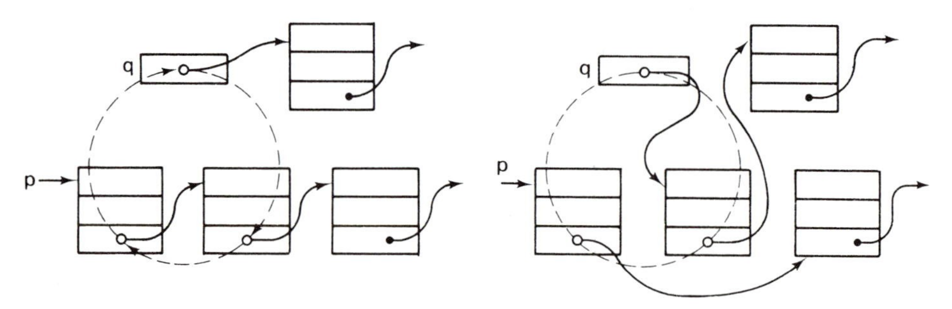

Fig. 4.9 List deletion and re-insertion.

The removal of a designated element itself (instead of its successor) is more difficult, because we encounter the same problem as with insertion: tracing backward to the denoted element's predecessor is impossible. But deleting the successor after moving its value forward is a relatively obvious and simple solution. It can be applied whenever p↑ has a successor, i.e., is not the last element on the list.

We now turn to the fundamental operation of *list traversal.* Let us assume that an operation P(x) has to be performed for every element of the list whose first element is p↑. This task is expressible as follows:

```
WHILE list designated by p is not empty DO
  perform operation P;
  proceed to the successor
END
```

In detail, this operation is descibed by statement (4.17).

```
WHILE p # NIL DO
  P(p↑); p := p↑.next                                    (4.17)
END
```

It follows from the definitions of the while statement and of the linking structure that P is applied to all elements of the list and to no other ones.

A very frequent operation performed is *list searching* for an element with a given key x. Unlike for arrays, the search must here be purely sequential. The search terminates either if an element is found or if the end of the list is reached. This is reflected by a logical conjunction consisting of two terms. Again, we assume that the head of the list is designated by a pointer p.

```
WHILE (p # NIL) & (p↑.key # x) DO p := p↑.next END       (4.18)
```

p = NIL implies that p↑ does not exist, and hence that the expression p↑.key # x is undefined. The order of the two terms is therefore essential.

4.3.2. Ordered Lists and Reorganizing Lists

Algorithm (4.18) strongly resembles the search routines for scanning an array or a sequence. In fact, a sequence is precisely a linear list for which the technique of linkage to the successor is left unspecified or implicit. Since the primitive sequence operators do not allow insertion of new elements (except at the end) or deletion (except removal of all elements), the choice of representation is left wide open to the implementor, and he may well use sequential allocation, leaving successive components in contiguous storage areas. Linear lists with explicit pointers provide more flexibility, and therefore they should be used whenever this additional flexibility is needed.

To exemplify, we will now consider a problem that will occur throughout this chapter in order to illustate alternative solutions and techniques. It is the problem of reading a text, collecting all its words, and counting the frequency of their occurrence. It is called the construction of a *concordance* or the generation of a *cross-reference list.*

An obvious solution is to construct a list of words found in the text. The list is scanned for each word. If the word is found, its frequency count is incremented; otherwise the word is added to the list. We shall simply call this process *search,* although it may actually also include an insertion. In order to be able to concentrate our attention on the essential part of list handling, we assume that the words have already been extracted from the text under investigation, have been encoded as integers, and are available in the from of an input sequence.

The formulation of the procedure called *search* follows in a straightforward manner from (4.18). The variable *root* refers to the head of the list in which new words are inserted according to (4.12). The complete algorithm is listed as Program 4.1; it includes a routine for tabulating the constructed cross-reference list. The tabulation process is an example in which an action is executed once for each element of the list, as shown in schematic form in (4.17).

```
MODULE List; (*straight list insertion*)
  FROM InOut IMPORT ReadInt, Done, WriteInt, WriteLn;
  FROM Storage IMPORT Allocate;

  TYPE Ptr = POINTER TO Word;
    Word =
      RECORD key: INTEGER;
        count: CARDINAL;
        next: Ptr
      END ;
  VAR k: INTEGER; root: Ptr;

  PROCEDURE search(x: INTEGER; VAR root: Ptr);
    VAR w: Ptr;
  BEGIN w := root;
    WHILE (w # NIL) & (w↑.key # x) DO w := w↑.next END ;
    (* (w = NIL) OR (w↑.key = x) *)
    IF w = NIL THEN (*new entry*)
      w := root; Allocate(root, SIZE(Word));
      WITH root↑ DO
        key := x; count := 1; next := w
      END
    ELSE w↑.count := w↑.count + 1
    END
  END search;

  PROCEDURE PrintList(w: Ptr);
  BEGIN
    WHILE w # NIL DO
      WriteInt(w↑.key, 8); WriteInt(w↑.count, 8); WriteLn;
      w := w↑.next
    END
  END PrintList;

BEGIN root := NIL; ReadInt(k);
  WHILE Done DO
    search(k, root); ReadInt(k)
  END ;
  PrintList(root)
END List.
```

Program 4.1 Straight List Insertion.

The linear scan algorithm of Program 4.1 resembles the search procedure for arrays, and reminds us of a simple technique used to simplify the loop termination condition: the use of a sentinel. A sentinel may as well be used in list search; it is represented by a dummy element at the end of the list. The new procedure is (4.20), which replaces the search procedure of Program 4.1, provided that a global variable *sentinel* is added and that the initialization of *root* is replaced by the statements

```
Allocate(sentinel, SIZE(Node));  root := sentinel                    (4.19)
```

which generate the element to be used as sentinel.

```
PROCEDURE search(x: INTEGER; VAR root: Ptr);                          (4.20)
  VAR w: Ptr;
BEGIN w := root; sentinel↑.key := x;
  WHILE w↑.key # x DO w := w↑.next END ;
  IF w = sentinel THEN  (*new entry*)
    w := root; Allocate(root, SIZE(Node));
    WITH root↑ DO
      key := x; count := 1; next := w
    END
  ELSE w↑.count := w↑.count + 1
  END
END search
```

Obviously, the power and flexibility of the linked list are ill used in this example, and the linear scan of the entire list can only be accepted in cases in which the number of elements is limited. An easy improvement, however, is readily at hand: the *ordered list search*. If the list is ordered (say by increasing keys), then the search may be terminated at the latest upon encountering the first key that is larger than the new one. Ordering of the list is achieved by inserting new elements at the appropriate place instead of at the head. In effect, ordering is practically obtained free of charge. This is because of the ease by which insertion in a linked list is achieved, i.e., by making full use of its flexibility. It is a possibility not provided by the array and sequence structures. (Note, however, that even in ordered lists no equivalent to the binary search of arrays is available.)

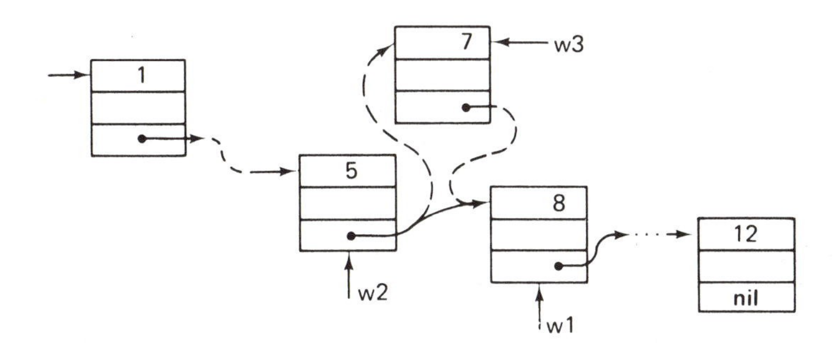

Fig. 4.10 Ordered list insertion.

Ordered list search is a typical example of the situation described in (4.15) in which an element must be inserted ahead of a given item, namely, in front of the first one whose key is too large. The technique shown here, however, differs from the one used in (4.15). Instead of copying values, *two* pointers are carried along in the list traversal; w2 lags one step behind w1 and thus identifies the proper insertion place when w1 has found too large a key. The general insertion step is shown in Fig. 4.10. The pointer to the new element (w3) is to be assigned to *w2↑.next,* except when the list is still empty. For reasons of simplicity and effectiveness, we prefer to avoid this distinction by using a conditional statement. The only way to avoid this is to introduce a dummy element at the list head. The initializing statement *root := NIL* in Program 4.1 is accordingly replaced by

```
Allocate(root, SIZE(Node)); root↑.next := NIL                    (4.21)
```

Referring to Fig. 4.10, we determine the condition under which the scan continues to proceed to the next element; it consists of two factors, namely,

```
(w1 # NIL) & (w1↑.key < x)                                       (4.22)
```

The resulting search procedure is shown in (4.23).

```
PROCEDURE search(x: INTEGER); VAR root: Ptr);                    (4.23)
  VAR w1, w2, w3: Ptr;
BEGIN (*w2 # NIL*)
  w2 := root; w1 := w2↑.next;
  WHILE (w1 # NIL) & (w1↑.key < x) DO
    w2 := w1; w1 := w2↑.next
  END ;
  (* (w1 = NIL) OR (w1↑.key >= x) *)
  IF (w1 = NIL) OR (w1↑.key > x) THEN (*new entry*)
    Allocate(w3, SIZE(Node)); w2↑.next := w3;
    WITH w3↑ DO
      key := x; count := 1; next := w1
    END
  ELSE w1↑.count := w1↑.count + 1
  END
END search
```

In order to speed up the search, the continuation condition of the while statement can once again be simplified by using a sentinel. This requires the initial presence of a dummy header as well as a sentinel at the tail.

It is now high time to ask what gain can be expected from ordered list search. Remembering that the additional complexity incurred is small, one should not expect an overwhelming improvement.

Assume that all words in the text occur with equal frequency. In this case the gain through lexicographical ordering is indeed also nil, once all words are listed, because the position of a word does not matter if only the total of all access steps is significant and if all words have the same frequency of occurrence. However, a gain is obtained whenever a new word is to be inserted. Instead of first scanning the entire list, on the average only half the list is scanned. Hence, ordered list insertion pays off only if a concordance is to be generated with many distinct words compared to their frequency of occurrence. The preceding examples are therefore suitable primarily as programming exercises rather than for practical applications.

The arrangement of data in a linked list is recommended when the number of elements is relatively small (< 50), varies, and, moreover, when no information is given about their frequencies of access. A typical example is the symbol table in compilers of programming languages. Each declaration causes the addition of a new symbol, and upon exit from its scope of validity, it is deleted from the list. The use of simple linked lists is appropriate for applications with relatively short programs. Even in this case a considerable improvement in access method can be achieved by a very simple technique which is mentioned here again primarily because it constitutes a pretty example for demonstrating the flexibilities of the linked list structure.

A characteristic property of programs is that occurrences of the same identifier are very often clustered, that is, one occurrence is often followed by one or more reoccurrences of the same word. This information is an invitation to reorganize the list after each access by moving the word that was found to the top of the list, thereby minimizing the length of the search path the next time it is sought. This method of access is called *list search with reordering,* or -- somewhat pompously -- self-organizing list search. In presenting the corresponding algorithm in the form of a procedure which may be substituted in Program 4.1, we take advantage of our experience made so far and introduce a sentinel right from the start. In fact, a sentinel not only speeds up the search, but in this case it also simplifies the program. The list must initially not be empty, but contains the sentinel element already. The initialization statements are

```
Allocate(sentinel, SIZE(Node)); root := sentinel;                    (4.24)
```

Note that the main difference between the new algorithm and the straight list search (4.21) is the action of reordering when an element has been found. It is then detached or deleted from its old position and inserted at the top. This deletion again requires the use of two chasing pointers, such that the predecessor w2 of an identified element w1 is still locatable. This, in turn, calls for the special treatment of the first element (i.e., the empty list). To conceive the linking process, we refer to Fig. 4.11. It shows the two pointers when w1 was identified as the desired element. The configuration after correct reordering is represented in Fig. 4.12, and the complete new search procedure is shown in (4.25).

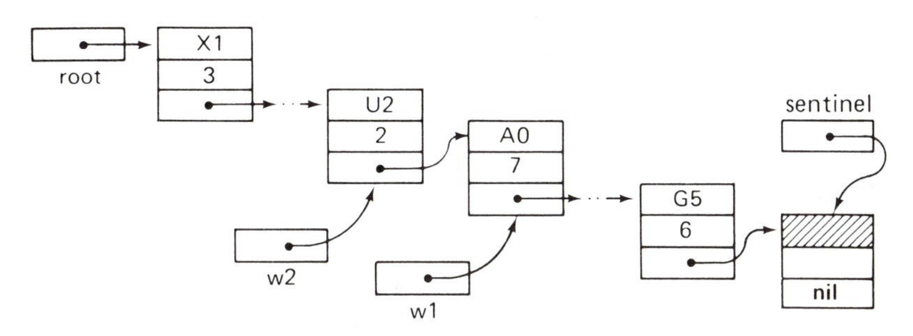

Fig. 4.11 List before reordering.

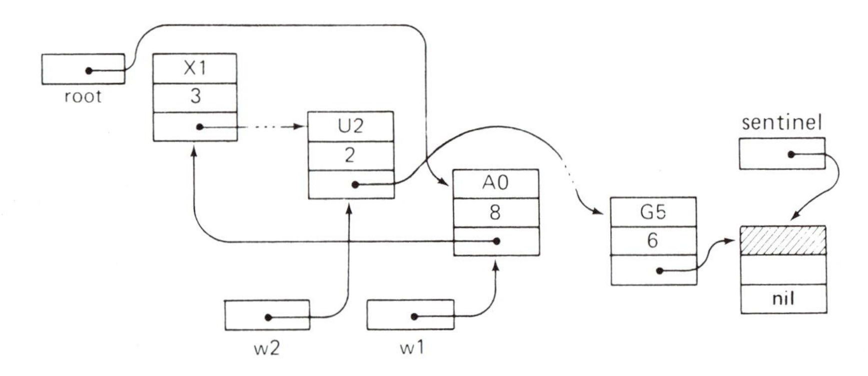

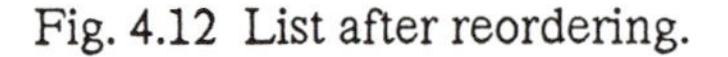
Fig. 4.12 List after reordering.

```
PROCEDURE search(x: integer; VAR root: Ptr);
  VAR w1, w2: Ptr;                                                    (4.25)
BEGIN w1 := root; sentinel↑.key := x;
  IF w1 = sentinel THEN (*first element*)
    Allocate(root, SIZE(Node));
    WITH root↑ DO
      key := x; count := 1; next := sentinel
    END
  ELSIF w1↑.key = x THEN w1↑.count := w1↑count + 1
  ELSE (*search*)
    REPEAT w2 := w1; w1 := w2↑.next
    UNTIL w1↑.key = x;
    IF w1 = sentinel THEN (*new entry*)
      w2 := root; Allocate(root, SIZE(Node));
      WITH root↑ DO
        key := x; count := 1; next := w2
      END
    ELSE (*found, now reorder*)
      w1↑.count := w1↑.count + 1;
      w2↑.next := w1↑.next; w1↑.next := root; root := w1
    END
  END
END search
```

The improvement in this search method strongly depends on the degree of clustering in the input data. For a given factor of clustering, the improvement will be more pronounced for large lists. To provide an idea of how much gain can be expected, an empirical measurement was made by applying the above cross-reference program to a short and a relatively long text and then comparing the methods of linear list ordering (4.20) and of list reorganization (4.25). The measured data are condensed into Table 4.2. Unfortunately, the improvement is greatest when a different data organization is needed anyway. We will return to this example in Sect. 4.4.

	Test 1	Test 2
Number of distinct keys	53	582
Number of occurrences of keys	315	14341
Time for search with ordering	6207	3200622
Time for search with reordering	4529	681584
Improvement factor	1.37	4.70

Table 4.2 Comparsion of List Search Methods.

4.3.3. An Application: Topological Sorting

An appropriate example of the use of a flexible, dynamic data structure is the process of *topological sorting.* This is a sorting process of items over which a *partial ordering* is defined, i.e., where an ordering is given over some pairs of items but not between all of them. The following are examples of partial orderings:

1. In a dictionary or glossary, words are defined in terms of other words. If a word v is defined in terms of a word w, we denote this by v ‹ w. Topological sorting of the words in a dictionary means arranging them in an order such that there will be no forward references.

2. A task (e.g., an engineering project) is broken up into subtasks. Completion of certain subtasks must usually precede the execution of other subtasks. If a subtask v must precede a subtask w, we write v ‹ w. Topological sorting means their arrangement in an order such that upon initiation of each subtask all its prerequisite subtasks have been completed.

3. In a university curriculum, certain courses must be taken before others since they rely on the material presented in their prerequisites. If a course v is a prerequisite for course w, we write v ‹ w. Topological sorting means arranging the courses in such an order that no course lists a later course as prerequisite.

4. In a program, some procedures may contain calls of other procedures. If a procedure v is called by a procedure w, we write v ‹ w. Topological sorting implies the arrangement of procedure declarations in such a way that there are no forward references.

In general, a partial ordering of a set S is a relation between the elements of S. It is denoted by the symbol ‹, verbalized by *precedes,* and satisfies the following three properties (axioms) for any distinct elements x, y, z of S:

(1) if x ‹ y and y ‹ z, then x ‹ z (transitivity)
(2) if x ‹ y, then not y ‹ z (asymmetry) (4.26)
(3) not z ‹ z (irreflexivity)

For evident reasons, we will assume that the sets S to be topologically sorted by an algorithm are finite. Hence, a partial ordering can be illustrated by drawing a diagram or graph in which the vertices denote the elements of S and the directed edges represent ordering relationships. An example is shown in Fig. 4.13.

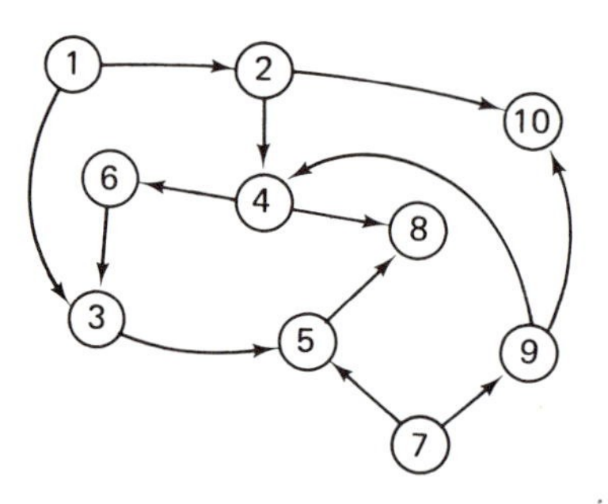

Fig. 4.13 Partially ordered set.

The problem of topological sorting is to embed the partial order in a linear order. Graphically, this implies the arrangement of the vertices of the graph in a row, such that all arrows point to the right, as shown in Fig. 4.14. Properties (1) and (2) of partial orderings ensure that the graph contains no loops. This is exactly the prerequisite condition under which such an embedding in a linear order is possible.

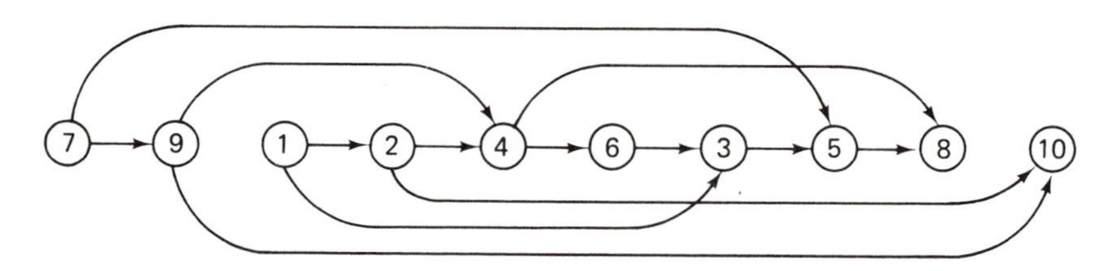

Fig. 4.14 Linear arrangement of partially ordered set of Fig. 4.13.

How do we proceed to find one of the possible linear orderings? The recipe is quite simple. We start by choosing any item that is not preceded by another item (there must be at least one; otherwise a loop would exist). This object is placed at the head of the resulting list and removed from the set S. The remaining set is still partially ordered, and so the same algorithm can be applied again until the set is empty.

In order to describe this algorithm more rigorously, we must settle on a data structure and representation of S and its ordering. The choice of this representation is determined by the operations to be performed, particularly the operation of selecting elements with zero predecessors. Every item should therefore be represented by three characteristics: its identification key, its set of successors, and a count of its predecessors. Since the number n of elements in S is not given *a priori*, the set is conveniently organized as a linked list. Consequently, an additional entry in the description of each item contains the link to the next item in the list. We will assume that the keys are integers (but not necessarily the consecutive integers from 1 to n). Analogously, the set of each item's successors is conveniently represented as a linked list. Each element of the successor list is described by an identification and a link to the next item on this list. If we call the descriptors of the main list, in which each item of S occurs exactly once, *leaders*, and the descriptors of elements on the successor chains *trailers*, we obtain the following declarations of data types:

```
TYPE LPtr =   POINTER TO leader;                                  (4.27)
     TPtr =   POINTER TO trailer;

     leader = RECORD key, count: INTEGER;
                trail: TPtr;  next: LPtr                          (4.28)
              END;

     trailer = RECORD id: LPtr;  next: TPtr
              END
```

Assume that the set S and its ordering relations are initially represented as a sequence of pairs of keys in the input file. The input data for the example in Fig. 4.13 are shown in (4.29) in which the symbols < are added for the sake of clarity.

```
1 < 2    2 < 4    4 < 6    2 < 10   4 < 8    6 < 3    1 < 3          (4.29)
3 < 5    5 < 8    7 < 5    7 < 9    9 < 4    9 < 10
```

The first part of the topological sort program must read the input and transform the data into a list structure. This is performed by successively reading a pair of keys x and y (x < y). Let us denote the pointers to their representations on the linked list of leaders by p and q. These records must be located by a list search and, if not yet present, be inserted in the list. This task is perfomed by a function procedure called *find*. Subsequently, a new entry is added in the list of trailers of x, along with an identification of y; the count of predecessors of y is incremented by 1. This algorithm is called *input phase* (4.30). Figure 4.15 illustrates the data structure generated during processing the input data (4.29) by (4.30). The function *find(w)* yields the reference to the list component with key w (see also Program 4.2).

```
(*input phase*)
Allocate(head, SIZE(leader)); tail := head; z := 0; ReadInt(x);       (4.30)
WHILE Done DO
  ReadInt(y); p := find(x); q := find(y);
  Allocate(t, SIZE(trailer)); t↑.id := q; t↑.next := p↑.trail;
  p↑.trail := t; q↑.count := q↑.count + 1; ReadInt(x)
END
```

After the data structure of Fig. 4.15 has been constructed in this input phase, the actual process of topological sorting can be taken up as described above. But since it consists of repeatedly selecting an element with a zero count of predecessors, it seems sensible to first gather all such elements in a linked chain. Since we note that the original chain of leaders will afterwards no longer be needed, the same field called *next* may be used again to link the zero predecessor leaders. This operation of replacing one chain by another chain occurs frequently in list processing. It is expressed in detail in (4.31), and for reasons of convenience it constructs the new chain in reverse order.

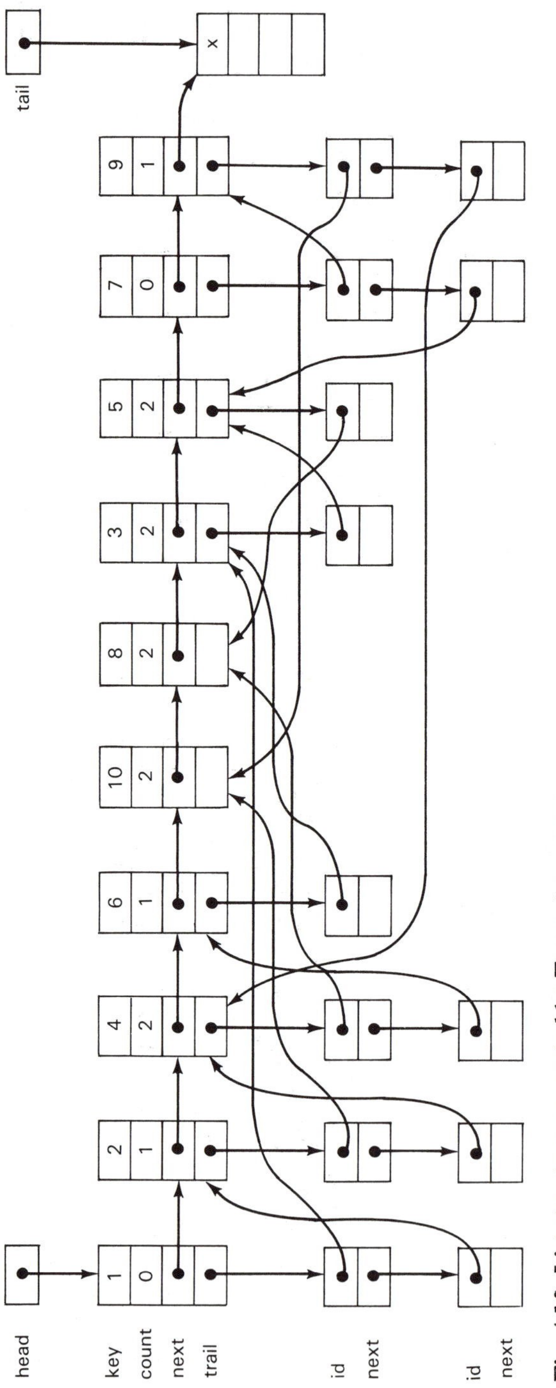

Fig. 4.15 List structure generated by Topsort program.

```
(*search for leaders without predecessors*)
p := head; head := NIL;
WHILE p # tail DO
  q := p; p := q↑.next;                                        (4.31)
  IF q↑.count = 0 THEN (*insert q↑ in new chain*)
    q↑.next := head; head := q
  END
END
```

Referring to Fig. 4.15, we see that the next chain of leaders is replaced by the one of Fig. 4.16 in which the pointers not depicted are left unchanged.

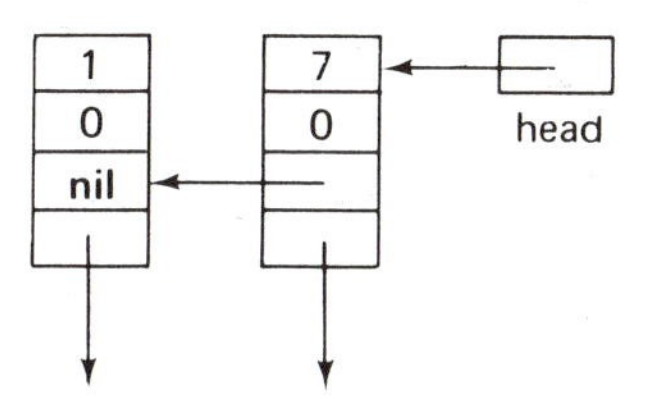

Fig. 4.16 List of leaders with zero counts.

After all this preparatory establishing of a convenient representation of the partially ordered set S, we can finally proceed to the actual task of topological sorting, i.e., of generating the output sequence. In a first rough version it can be described as follows:

```
q := head;
WHILE q # NIL DO (*output this element, then delete it*)
  WriteInt(q↑.key, 8); n := n - 1;
  t := q↑.trail; q := q↑.next;                                 (4.32)
  decrement the predecessor count of all its successors
  on trailer list t; if any count becomes 0, insert this
  element in the leader list q
END
```

The statement in (4.32) that is to be still further refined constitutes one more scan of a list [see schema (4.17)]. In each step, the auxiliary variable p designates the leader element whose count has to be decremented and tested.

```
WHILE t # NIL DO
  p := t↑.id; p↑.count := p↑.count - 1;                        (4.33)
  IF p↑.count = 0 THEN (*insert p↑ in leader list*)
    p↑.next := q; q := p
  END ;
  t := t↑.next
END
```

This completes the program for topological sorting. Note that a counter n was introduced to count the leaders generated in the input phase. This count is decremented each time a leader element is output in the output phase. It should therefore return to zero at the end of the program. Its failure to return to zero is an indication that there are elements left in the structure when none is without predecessor. In this case the set S is evidently not partially ordered. The output phase programmed above is an example of a process that maintains a list that pulsates, i.e., in which elements are inserted and removed in an unpredictable order. It is therefore an example of a process which utilizes the full flexibility offered by the explicity linked list.

```
MODULE TopSort;
  FROM InOut IMPORT OpenInput, CloseInput,
     ReadInt, Done, WriteInt, WriteString, WriteLn;
  FROM Storage IMPORT ALLOCATE;

  TYPE LPtr = POINTER TO leader;
    TPtr =      POINTER TO trailer;

    leader =    RECORD key, count: INTEGER;
                  trail: TPtr; next: LPtr
                END ;

    trailer =   RECORD id: LPtr;  next: TPtr
                END ;

  VAR p, q, head, tail: LPtr;
    t: TPtr;
    x, y, n: INTEGER;

  PROCEDURE find(w: INTEGER): LPtr;
    VAR h: LPtr;
  BEGIN h := head; tail↑.key := w;  (*sentinel*)
    WHILE h↑.key # w DO h := h↑.next END ;
    IF h = tail THEN
      ALLOCATE(tail, SIZE(leader)); n := n+1;
      h↑.count := 0; h↑.trail := NIL; h↑.next := tail
    END ;
    RETURN h
  END find;

BEGIN
  (*initialize list of leaders with a dummy acting as sentinel*)
  ALLOCATE(head, SIZE(leader)); tail := head; n := 0;

  OpenInput("TEXT"); ReadInt(x);
  WHILE Done DO
    WriteInt(x, 8); ReadInt(y); WriteInt(y, 8); WriteLn;
    p := find(x); q := find(y);
    ALLOCATE(t, SIZE(trailer)); t↑.id := q; t↑.next := p↑.trail;
    p↑.trail := t; q↑.count := q↑.count + 1; ReadInt(x)
```

```
  END ;
  CloseInput;

  (*search for leaders without predecessors*)
  p := head; head := NIL;
  WHILE p # tail DO
    q := p; p := q↑.next;
    IF q↑.count = 0 THEN (*insert q↑ in new chain*)
      q↑.next := head; head := q
    END
  END ;

  (*output phase*) q := head;
  WHILE q # NIL DO
    WriteLn; WriteInt(q↑.key, 8); n := n-1;
    t := q↑.trail; q := q↑.next;
    WHILE t # NIL DO
      p := t↑.id; p↑.count := p↑.count - 1;
      IF p↑.count = 0 THEN (*insert p↑ in leader list*)
        p↑.next := q; q := p
      END ;
      t := t↑.next
    END
  END ;
  IF n # 0 THEN WriteString("This set is not partially ordered") END ;
  WriteLn
END TopSort.
```

Program 4.2 Topological Sort

4.4 TREE STRUCTURES

4.4.1. Basic Concepts and Definitions

We have seen that sequences and lists may conveniently be defined in the following way: A sequence (list) with base type T is either

1. The empty sequence (list).
2. The concatenation (chain) of a T and a sequence with base type T.

Hereby recursion is used as an aid in defining a structuring principle, namely, sequencing or iteration. Sequences and iterations are so common that they are usually considered as fundamental patterns of structure and behaviour. But it should be kept in mind that they can be defined in terms of recursion, whereas the reverse is not true, for recursion may be effectively and elegantly used to define much more sophisticated structures. Trees are a well-known example. Let a tree structure be defined as follows: A *tree structure* with base type T is either

1. The empty structure.
2. A node of type T with a finite number of associated disjoint tree structures of base type T, called *subtrees*.

From the similarity of the recursive definitions of sequences and tree structures it is evident that the sequence (list) is a tree structure in which each node has at most one subtree. The sequence (list) is therefore also called a *degenerate* tree.

There are several ways to represent a tree structure. For example, a tree structure with its base type T ranging over the letters is shown in various ways in Fig. 4.17. These representations all show the same structure and are therefore equivalent. It is the graph structure that explicitly illustrates the branching relationships which, for obvious reasons, led to the generally used name *tree*. Strangely enough, it is customary to depict trees upside down, or -- if one prefers to express this fact differently -- to show the roots of trees. The latter formulation, however, is misleading, since the top node (A) is commonly called the root.

An *ordered tree* is a tree in which the branches of each node are ordered. Hence the two ordered trees in Fig. 4.18 are distinct, different objects. A node y that is directly below node x is called a (direct) *descendant* of x; if x is at level i, then y is said to be a level i+1. Conversely, node x is said to be the (direct) *ancestor* of y. The root of a tree is defined to be at level 0. The maximum level of any element of a tree is said to be its *depth* or *height*.

If an element has no descendants, it is called a *terminal* node or a *leaf*; and an element that is not terminal is an *interior* node. The number of (direct) descendants of an interior node is called its *degree*. The maximum degree over all nodes is the degree of the tree. The number of branches or edges that have to be traversed in order to proceed from the root to a node x is called the *path length* of x. The root has path length 0, its direct descendants have path length 1, etc. In general, a node at level i has path length i. The path length of a tree is

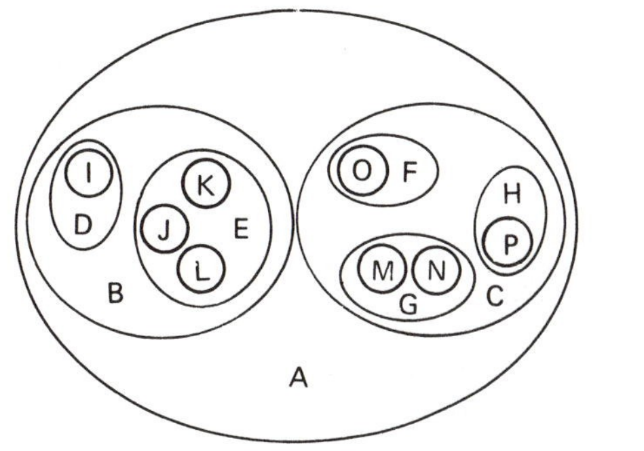

(a)

(A (B (D (I), E (J, K, L)), C (F (O), G (M, N), H (P)))) (b)

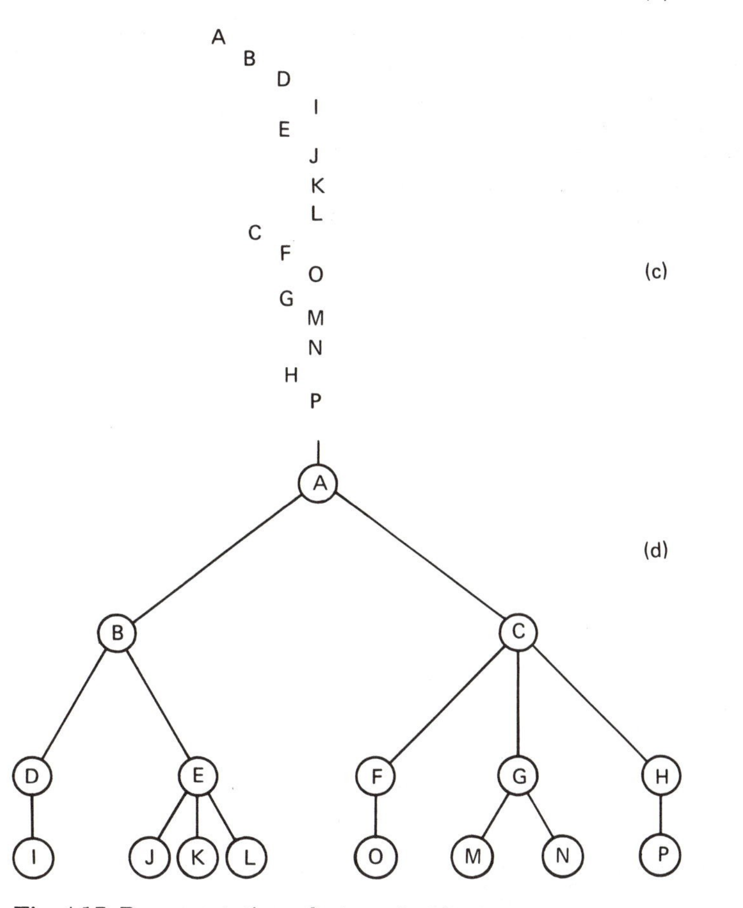

Fig. 4.17 Representation of a tree structure:
(a) Nested sets; (b) nested parentheses; (c) indentation; (d) graph.

defined as the sum of the path lengths of all its components. It is also called its *internal path length*. The internal path length of the tree shown in Fig. 4.17, for instance, is 36. Evidently, the average path length is

$$P_I = (Si: 1 \leq i \leq n : n_i * i) / n \qquad (4.34)$$

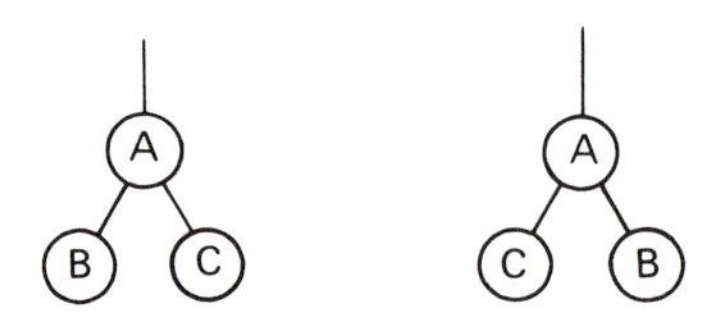

Fig. 4.18 Two distinct binary trees.

where n_i is the number of nodes at level i. In order to define what is called the *external path length*, we extend the tree by a special node wherever a subtree was missing in the original tree. In doing so, we assume that all nodes are to have the same degree, namely the degree of the tree. Extending the tree in this way therefore amounts to filling up empty branches, whereby the special nodes, of course, have no further descendants. The tree of Fig. 4.17 extended with special nodes is shown in Fig. 4.19 in which the special nodes are represented by squares. The external path length is now defined as the sum of the path lengths over all special nodes. If the number of special nodes at level i is m_i, then the average external path length is

$$P_E = (\mathbf{S}i: 1 \leq i \leq m : m_i * i) / m \qquad (4.35)$$

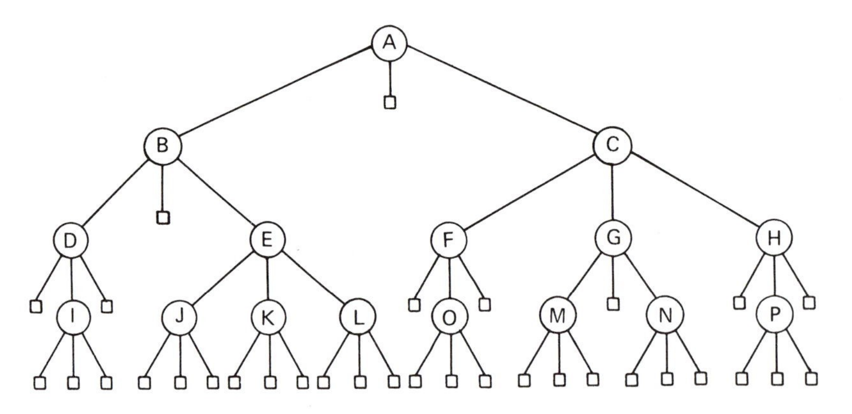

Fig. 4.19 Ternary tree extended with special nodes.

In the tree shown in Fig. 4.19 the external path length is 120. The number of special nodes m to be added in a tree of degree d directly depends on the number n of original nodes. Note that every node has exactly one edge pointing to it. Thus, there are m+n edges in the extended tree. On the other hand, d edges are emanating from each original node, none from the special nodes. Therefore, there exist d*n + 1 edges, the 1 resulting from the edge pointing to the root. The two results yield the following equation between the number m of special nodes and n of original nodes: d*n + 1 = m+n, or

$$m = (d-1)*n + 1 \qquad (4.36)$$

The maximum number of nodes in a tree of a given height h is reached if all nodes have d subtrees, except those at level h, all of which have none. For a tree of degree d, level 0 then contains 1 node (namely, the root), level 1 contains its d descendants, level 2 contains the d^2 descendants of the d nodes at level 2, etc. This yields

$$N_d(h) \;=\; \mathbf{S}i\colon 0 \le i < h : d^i \tag{4.37}$$

as the maximum number of nodes for a tree with height h and degree d. For d = 2, we obtain

$$N_2(h) \;=\; 2^h - 1 \tag{4.38}$$

Of particular importance are the ordered trees of degree 2. They are called *binary trees*. We define an ordered binary tree as a finite set of elements (nodes) which either is empty or consists of a root (node) with two disjoint binary trees called the *left* and the *right subtree* of the root. In the following sections we shall exclusively deal with binary trees, and we therefore shall use the word tree to mean *ordered binary tree*. Trees with degree greater than 2 are called *multiway trees* and are discussed in Sect. 7 of this chapter.

Familiar examples of binary trees are the family tree (pedigree) with a person's father and mother as descendants (!), the history of a tennis tournament with each game being a node denoted by its winner and the two previous games of the combatants as its descendants, or an arithmetic expression with dyadic operators, with each operator denoting a branch node with its operands as subtrees (see Fig. 4.20).

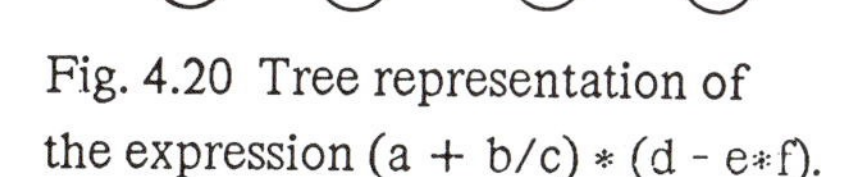

Fig. 4.20 Tree representation of the expression (a + b/c) * (d - e*f).

We now turn to the problem of representation of trees. It is plain that the illustration of such recursive structures in terms of branching structures immediately suggests the use of our pointer facility. There is evidently no use in declaring variables with a fixed tree structure; instead, we define the nodes as variables with a fixed structure, i.e., of a fixed type, in which the degree of the tree determines the number of pointer components referring to the node's subtrees. Evidently, the reference to the empty tree is denoted by NIL. Hence, the tree of Fig. 4.20 consists of components of a type defined as follows and may then be constructed as shown in Fig. 4.21.

```
TYPE Ptr =    POINTER TO Node;
TYPE Node = RECORD op: CHAR;                                   (4.39)
                left, right: Ptr
              END
```

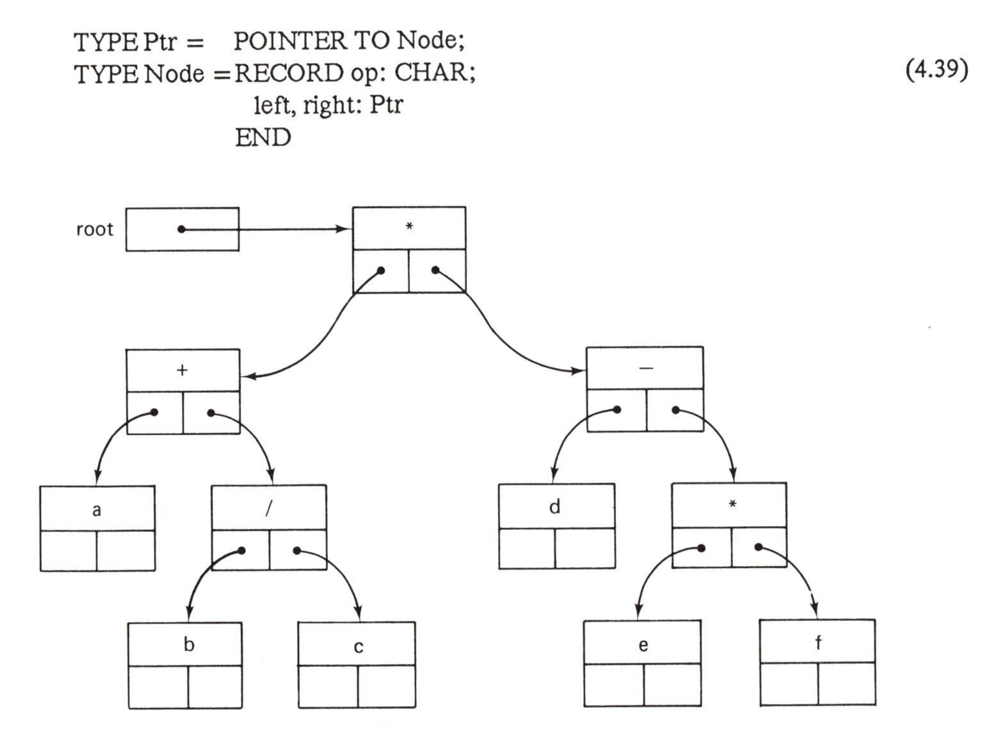

Fig. 4.21 Tree represented as data structure.

Before investigating how trees might be used advantageously and how to perform operations on trees, we give an example of how a tree may be constructed by a program. Assume that a tree is to be generated containing nodes of the type defined in (4.39), with the values of the nodes being n numbers read from an input file. In order to make the problem more challenging, let the task be the construction of a tree with n nodes and minimal height. In order to obtain a minimal height for a given number of nodes, one has to allocate the maximum possible number of nodes of all levels except the lowest one. This can clearly be achieved by distributing incoming nodes equally to the left and right at each node. This implies that we structure the tree for given n as shown in Fig. 4.22, for n = 1, ... , 7.

The rule of equal distribution under a known number n of nodes is best formulated recursively:

1. Use one node for the root.
2. Generate the left subtree with nl = n DIV 2 nodes in this way.
3. Generate the right subtree with nr = n - nl - 1 nodes in this way.

The rule is expressed as a recursive procedure as part of Program 4.3, which reads the input file and constructs the perfectly balanced tree. We note the following definition: A tree is *perfectly balanced,* if for each node the numbers of nodes in its left and right subtrees differ by at most 1.

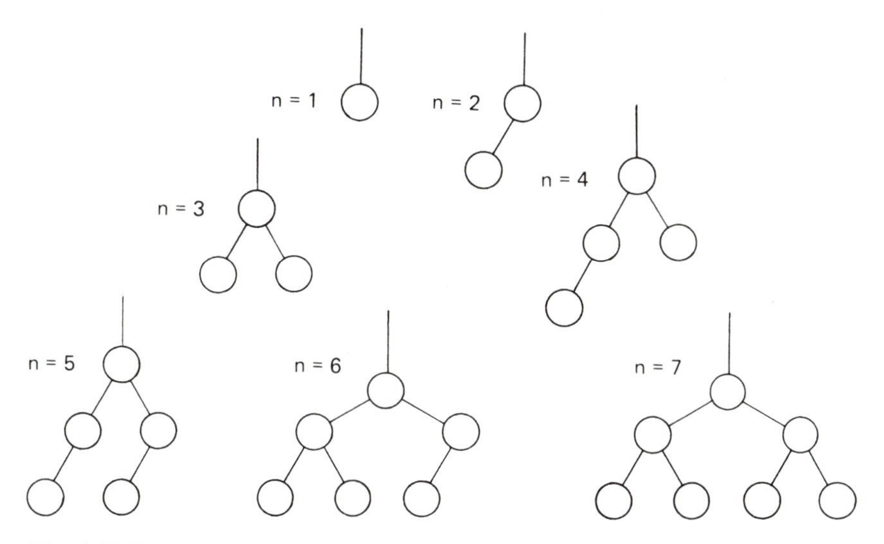

Fig. 4.22 Perfectly balanced trees.

```
MODULE BuildTree;
  FROM InOut IMPORT OpenInput, CloseInput,
     ReadInt, WriteInt, WriteString, WriteLn;
  FROM Storage IMPORT ALLOCATE;

  TYPE Ptr = POINTER TO Node;

  Node = RECORD key: INTEGER;
          left, right: Ptr
       END ;

  VAR n: INTEGER; root: Ptr;

  PROCEDURE tree(n: INTEGER): Ptr;
    VAR newnode: Ptr;
       x, nl, nr: INTEGER;
  BEGIN (*construct perfectly balanced tree with n nodes*)
    IF n = 0 THEN newnode := NIL
    ELSE nl := n DIV 2; nr := n-nl-1;
      ReadInt(x); ALLOCATE(newnode, SIZE(Node));
      WITH newnode↑ DO
        key := x; left := tree(nl); right := tree(nr)
      END
    END ;
    RETURN newnode
  END tree;
```

```
PROCEDURE PrintTree(t: Ptr; h: INTEGER);
  VAR i: INTEGER;
BEGIN (*print tree t with indentation h*)
  IF t # NIL THEN
    WITH t↑ DO
      PrintTree(left, h+1);
      FOR i := 1 TO h DO WriteString("     ") END ;
      WriteInt(key, 6); WriteLn;
      PrintTree(right, h+1)
    END
  END
END PrintTree;

BEGIN (*first integer is number of nodes*)
  OpenInput("TEXT"); ReadInt(n);
  root := tree(n);
  PrintTree(root, 0); CloseInput
END BuildTree.
```

Program 4.3 Construct Perfectly Balanced Tree.

Assume, for example, the following input data for a tree with 21 nodes:

21 8 9 11 15 19 20 21 7 3 2 1 5 6 4 13 14 10 12 17 16 18

Program 4.3 then constructs the perfectly balanced tree shown in Fig. 4.23. Note the simplicity and transparency of this program that is obtained through the use of recursive procedures. It is obvious that recursive algorithms are particularly suitable when a program is to manipulate information whose structure is itself defined recursively. This is again manifested in the procedure which prints the resulting tree: the empty tree results in no printing, the subtree at level L in first printing its own left subtree, then the node, properly indented by preceding it with L blanks, and finally in printing its right subtree.

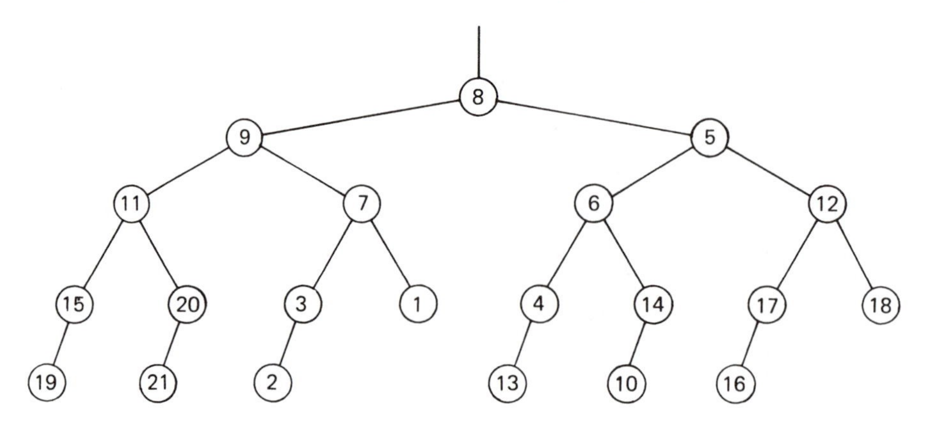

Fig. 4.23 Tree generated by Program 4.3.

4.4.2. Basic Operations on Binary Trees

There are many tasks that may have to be perfomed on a tree structure; a common one is that of executing a given operation P on each element of the tree. P is then understood to be a parameter of the more general task of visting all nodes or, as it is usually called, of *tree traversal*. If we consider the task as a single sequential process, then the individual nodes are visited in some specific order and may be considered as being laid out in a linear arrangement. In fact, the description of many algorithms is considerably facilitated if we can talk about processing the next element in the tree based in an underlying order. There are three principal orderings that emerge naturally from the structure of trees. Like the tree structure itself, they are conveniently expressed in recursive terms. Referring to the binary tree in Fig. 4.24 in which R denotes the root and A and B denote the left and right subtrees, the three orderings are

1. Preorder: R, A, B (visit root before the subtrees)
2. Inorder: A, R, B
3. Postorder: A, B, R (visit root after the subtrees)

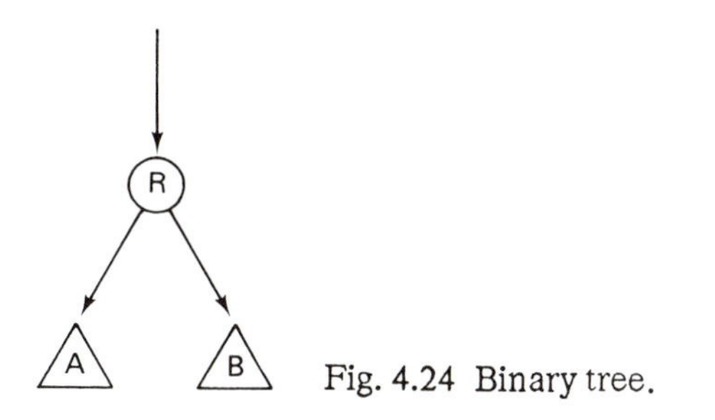

Fig. 4.24 Binary tree.

Traversing the tree of Fig. 4.20 and recording the characters seen at the nodes in the sequence of encounter, we obtain the following orderings:

1. Preorder: * + a / b c - d * e f
2. Inorder: a + b / c * d - e * f
3. Postorder: a b c / + d e f * - *

We recognize the three forms of expressions: *preorder* traversal of the expression tree yields *prefix* notation; *postorder* traversal generates *postfix* notation; and *inorder* traversal yields conventional *infix* notation, although without the parentheses necessary to clarify operator precedences.

Let us now formulate the three methods of traversal by three concrete programs with the explicit parameter t denoting the tree to be operated upon and with the implicit parameter P denoting the operation to be performed on each node. Assume the following definitions:

```
TYPE Ptr =   POINTER TO Node;
TYPE Node =RECORD ...                                   (4.42)
                 left, right: Ptr
```

```
                END
```

The three methods are now readily formulated as recursive procedures; they demonstrate again the fact that operations on recursively defined data structures are most conveniently defined as recursive algorithms.

```
PROCEDURE preorder(t: Ptr);
BEGIN                                                        (4.43)
  IF t # NIL THEN
    P(t); preorder(t↑.left); preorder(t↑.right)
  END
END preorder

PROCEDURE inorder(t: Ptr);
BEGIN                                                        (4.44)
  IF t # NIL THEN
    inorder(t↑.left); P(t); inorder(↑.right)
  END
END inorder

PROCEDURE postorder(t: Ptr);
BEGIN                                                        (4.45)
  IF t # NIL THEN
    postorder(t↑.left); postorder(t↑.right); P(t)
  END
END postorder
```

Note that the pointer t is passed as a value parameter. This expresses the fact that the relevant entity is the reference to the considered subtree and not the variable whose value is the pointer, and which could be changed in case t were passed as a variable parameter.

An example of a tree traversal routine is that of printing a tree, with appropriate indentation indicating each node's level (see Program 4.3).

Binary trees are frequently used to represent a set of data whose elements are to be retrievable through a unique key. If a tree is organized in such a way that for each node t_i, all keys in the left subtree of t_i are less than the key of t_i, and those in the right subtree are greater than the key of t_i, then this tree is called a *search tree*. In a search tree it is possible to locate an arbitrary key by starting at the root and proceeding along a search path switching to a node's left or right subtree by a decision based on inspection of that node's key only. As we have seen, n elements may be organized in a binary tree of a height as little as log n. Therefore, a search among n items may be performed with as few as log n comparsions if the tree is perfectly balanced. Obviously, the tree is a much more suitable form for organizing such a set of data than the linear list used in the previous section. As this search follows a single path from the root to the desired node, it can readily be programmed by iteration (4.46).

```
PROCEDURE locate(x: INTEGER; t: Ptr): Ptr;
```

```
BEGIN                                                          (4.46)
  WHILE (t # NIL) & (t↑.key # x) DO
    IF t↑.key < x THEN t := t↑.right ELSE t := t↑.left END
  END ;
  RETURN t
END locate
```

The function *locate(x, t)* yields the value NIL, if no key with value x is found in the tree with root t. As in the case of searching a list, the complexity of the termination condition suggests that a better solution may exist, namely the use of a sentinel. This technique is equally applicable in the case of a tree. The use of pointers makes it possible for all branches of the tree to terminate with the same sentinel. The resulting structure is no longer a tree, but rather a tree with all leaves tied down by strings to a single anchor point (Fig. 4.25). The sentinel may be considered as a common, shared representative of all external nodes by which the original tree was extended (see Fig. 4.19). The resulting, simplified search routine is listed in (4.47).

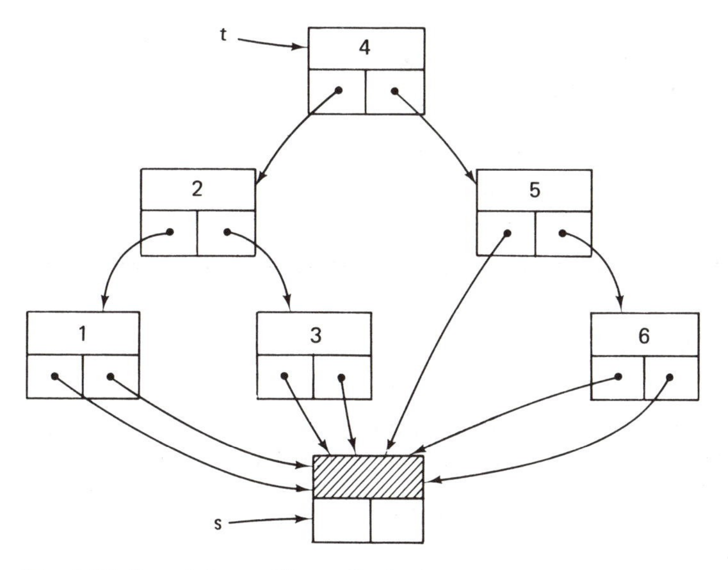

Fig. 4.25 Search tree with sentinel.

```
PROCEDURE locate(x: INTEGER; t: Ptr): Ptr;
BEGIN s↑.key := x; (*sentinel*)
  WHILE t↑.key # x DO                                          (4.47)
    IF t↑.key < x THEN t := t↑.right ELSE t := t↑.left END
  END ;
  RETURN t
END locate
```

Note that in this case *locate(x, t)* yields the value s instead of NIL, i.e., the pointer to the sentinel, if no key with value x is found in the tree with root t.

4.4.3. Tree Search and Insertion

The full power of the dynamic allocation technique with access through pointers is hardly displayed by those examples in which a given set of data is built, and thereafter kept unchanged. More suitable examples are those applications in which the structure of the tree itself varies, i.e., grows and/or shrinks during the execution of the program. This is also the case in which other data representations, such as the array, fail and in which the tree with elements linked by pointers emerges as the most appropriate solution.

We shall first consider only the case of a steadily growing but never shrinking tree. A typical example is the concordance problem which was already investigated in connection with linked lists. It is now to be revisited. In this problem a sequence of words is given, and the number of occurrences of each word has to be determined. This means that, starting with an empty tree, each word is searched in the tree. If it is found, its occurrence count is incremented; otherwise it is inserted as a new word (with a count initialized to 1). We call the underlying task *tree search with insertion*. The following data type definitions are assumed:

```
TYPE WPtr =POINTER TO Word;
  Word =       RECORD                                   (4.48)
                 key: INTEGER;
                 count: CARDINAL;
                 left, right: WPtr
               END
```

Assuming moreover a source of keys and a variable denoting the root of the search tree, we may formulate the program as

```
ReadInt(x);                                              (4.49)
WHILE Done DO search(x, root); ReadInt(x) END
```

Finding the search path is again straightforward. However, if it leads to a dead end (i.e., to an empty subtree designated by a pointer value NIL), then the given word must be inserted in the tree at the place of the empty subtree. Consider, for example, the binary tree shown in Fig. 4.26 and the insertion of the name *Paul*. The result is shown in dotted lines in the same picture.

The entire operation is shown in Program 4.4. The search process is formulated as a recursive procedure. Note that its parameter p is a variable parameter and not a value parameter. This is essential because in the case of insertion a new pointer value must be assigned to the variable which previously held the value NIL. Using the input sequence of 21 numbers that had been applied to Program 4.3 to construct the tree of Fig. 4.23, Program 4.4 yields the binary search tree shown in Fig. 4.27.

```
MODULE TreeSearch;
  FROM InOut IMPORT OpenInput, CloseInput,
     ReadInt, Done, WriteInt, WriteString, WriteLn;
```

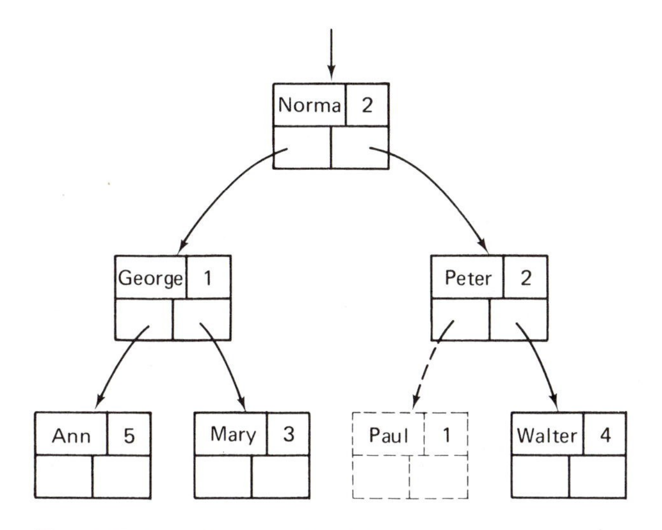

Fig. 4.26 Insertion in ordered binary tree.

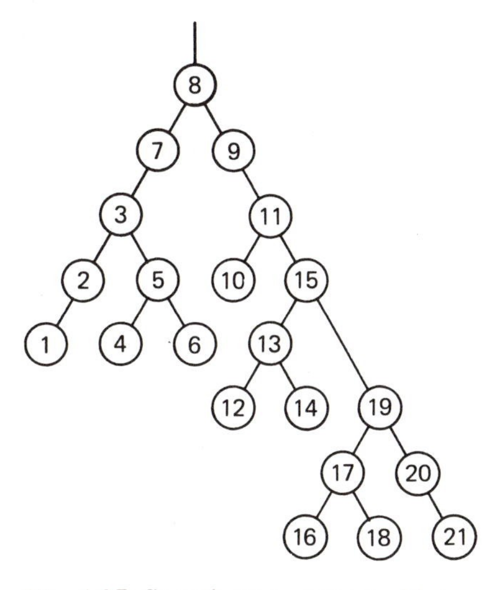

Fig. 4.27 Search tree generated by Program 4.4.

```
FROM Storage IMPORT ALLOCATE;

TYPE WPtr = POINTER TO Word;
   Word = RECORD key: INTEGER;
          count: INTEGER;
          left, right: WPtr
        END ;

VAR root: WPtr; n, key: INTEGER;
```

```
PROCEDURE PrintTree(t: WPtr; h: INTEGER);
  VAR i: INTEGER;
BEGIN (*print tree t with indentation h*)
  IF t # NIL THEN
    WITH t↑ DO
      PrintTree(left, h+1);
      FOR i := 1 TO h DO WriteString("     ") END ;
      WriteInt(key, 6); WriteLn;
      PrintTree(right, h+1)
    END
  END
END PrintTree;

PROCEDURE search(x: INTEGER; VAR p: WPtr);
BEGIN
  IF p = NIL THEN (*word not in tree; insert*)
    ALLOCATE(p, SIZE(Word));
    WITH p↑ DO
      key := x; count := 1; left := NIL; right := NIL
    END
  ELSIF x < p↑.key THEN search(x, p↑.left)
  ELSIF x > p↑.key THEN search(x, p↑.right)
  ELSE p↑.count := p↑.count + 1
  END
END search;

BEGIN root := NIL;
  (*first integer is number of nodes*)
  OpenInput("TEXT"); ReadInt(n);
  WHILE n > 0 DO
    ReadInt(key); search(key, root); n := n-1
  END ;
  CloseInput; PrintTree(root, 0)
END TreeSearch.
```

Program 4.4 Tree Search and Insertion.

The use of a sentinel again simplifies the task somewhat, as is shown in (4.50). Clearly, at the start of the program the variable *root* must be initialized by the pointer to the sentinel instead of the value NIL, and before each search the specified value x must be assigned to the key field of the sentinel.

```
PROCEDURE search(x: INTEGER; VAR p: WPtr);
BEGIN
  IF x < p↑.key THEN search(x, p↑.left)
  ELSIF x > p↑.key THEN search(x, p↑.right)
  ELSIF p # s THEN p↑.count := p↑.count + 1
```

```
    ELSE (*insert*) Allocate(p, SIZE(Word));
      WITH p↑ DO                                              (4.50)
        key := x; left := s; right := s; count := 1
      END
    END
  END
```

Although the purpose of this algorithm is searching, it can be used for sorting as well. In fact, it resembles the sorting by insertion method quite strongly, and because of the use of a tree structure instead of an array, the need for relocation of the components above the insertion point vanishes. Tree sorting can be programmed to be almost as efficient as the best array sorting methods known. But a few precautions must be taken. After encountering a match, the new element must also be inserted. If the case $x = p\uparrow.key$ is handled identically to the case $x > p\uparrow.key$, then the algorithm represents a stable sorting method, i.e., items with identical keys turn up in the same sequence when scanning the tree in normal order as when they were inserted.

In general, there are better ways to sort, but in applications in which searching and sorting are both needed, the tree search and insertion algorithm is strongly recommended. It is, in fact, very often applied in compilers and in data banks to organize the objects to be stored and retrieved. An appropriate example is the construction of a cross-reference index for a given text, an example that we had already used to illustrate list generation.

Our task is to construct a program that (while reading a text and printing it after supplying consecutive line numbers) collects all words of this text, thereby retaining the numbers of the lines in which each word occurred. When this scan is terminated, a table is to be generated containing all collected words in alphabetical order with lists of their occurrences.

Obviously, the search tree (also called a lexicographic tree) is a most suitable candidate for representing the words encountered in the text. Each node now not only contains a word as key value, but it is also the head of a list of line numbers. We shall call each recording of an occurrence an *item*. Hence, we encounter both trees and linear lists in this example. The program consists of two main parts (see Program 4.5), namely, the scanning phase and the table printing phase. The latter is a straightforward application of a tree traversal routine in which visiting each node implies the printing of the key value (word) and the scanning of its associated list of line numbers (items). The following are further clarifications regarding the Cross Reference Generator of Program 4.5. Table 4.4 shows the results of processing the program text (4.50).

1. A word is considered as any sequence of letters and digits starting with a letter.

2. Since words may be of widely different lengths, the actual characters are stored in an array *buffer*, and the tree nodes contain the index of the key's first character.

3. It is desirable that the line numbers be printed in ascending order in the cross-reference index. Therefore, the item lists must be generated in the same order as they are scanned upon printing. This requirement suggests the use of two pointers in each word node, one

referring to the first, and one referring to the last item on the list.

```
MODULE CrossRef;
  FROM InOut IMPORT OpenInput, OpenOutput, CloseInput,
     CloseOutput, Read, Done, EOL, Write, WriteCard, WriteLn;
  FROM Storage IMPORT ALLOCATE;

  CONST BufLeng = 10000; WordLeng = 16;

  TYPE WordPtr = POINTER TO Word;
     ItemPtr = POINTER TO Item;

     Word = RECORD key: CARDINAL;
           first, last: ItemPtr;
           left, right: WordPtr
         END ;

     Item = RECORD lno: CARDINAL;
           next: ItemPtr
         END ;

  VAR  root: WordPtr;
     k0, k1, line: CARDINAL;
     ch: CHAR;
     buffer: ARRAY [0 .. BufLeng-1] OF CHAR;

  PROCEDURE PrintWord(k: CARDINAL);
    VAR lim: CARDINAL;
  BEGIN lim := k + WordLeng;
    WHILE buffer[k] > 0C DO Write(buffer[k]); k := k+1 END ;
    WHILE k < lim DO Write(" "); k := k+1 END
  END PrintWord;

  PROCEDURE PrintTree(t: WordPtr);
    VAR i, m: INTEGER; item: ItemPtr;
  BEGIN
    IF t # NIL THEN
      WITH t↑ DO
        PrintTree(left);
        PrintWord(key); item := first; m := 0;
        REPEAT
          IF m = 8 THEN
            WriteLn; m := 0;
            FOR i := 1 TO WordLeng DO Write(" ") END
          END ;
          m := m+1; WriteCard(item↑.lno, 6); item := item↑.next
        UNTIL item = NIL;
        WriteLn;
        PrintTree(right)
```

```
      END
    END
  END PrintTree;

  PROCEDURE Diff(i, j: CARDINAL): INTEGER;
  BEGIN
    LOOP
      IF buffer[i] # buffer[j] THEN
        RETURN INTEGER(ORD(buffer[i])) - INTEGER(ORD(buffer[j]))
      ELSIF buffer[i] = 0C THEN RETURN 0
      END ;
      i := i+1; j := j+1
    END
  END Diff;

  PROCEDURE search(VAR p: WordPtr);
    VAR item: ItemPtr; d: INTEGER;
  BEGIN
    IF p = NIL THEN (*word not in tree; insert*)
      ALLOCATE(p, SIZE(Word)); ALLOCATE(item, SIZE(Item));
      WITH p↑ DO
        key := k0; first := item; last := item;
        left := NIL; right := NIL
      END ;
      item↑.lno := line; item↑.next := NIL; k0 := k1
    ELSE d := Diff(k0, p↑.key);
      IF d < 0 THEN search(p↑.left)
      ELSIF d > 0 THEN search(p↑.right)
      ELSE ALLOCATE(item, SIZE(Item));
        item↑.lno := line; item↑.next := NIL;
        p↑.last↑.next := item; p↑.last := item
      END
    END
  END search;

  PROCEDURE GetWord;
  BEGIN k1 := k0;
    REPEAT Write(ch); buffer[k1] := ch; k1 := k1 + 1; Read(ch)
    UNTIL (ch < "0") OR (ch > "9") & (CAP(ch) < "A")
      OR (CAP(ch) > "Z");
    buffer[k1] := 0C; k1 := k1 + 1; (*terminator*)
    search(root)
  END GetWord;

BEGIN root := NIL; k0 := 0; line := 0;
  OpenInput("TEXT"); OpenOutput("XREF");
  WriteCard(0, 6); Write(" "); Read(ch);
```

```
  WHILE Done DO
    CASE ch OF
      0C .. 35C:     Read(ch) |
      36C .. 37C:    WriteLn; Read(ch); line := line + 1;
                     WriteCard(line, 6); Write(" ") |
      " " .. "@":    Write(ch); Read(ch) |
      "A" .. "Z":    GetWord |
      "[" .. "‘":    Write(ch); Read(ch) |
      "a" .. "z":    GetWord |
      "{" .. "~":    Write(ch); Read(ch)
    END
  END ;
  WriteLn; WriteLn; CloseInput;
  PrintTree(root); CloseOutput
END CrossRef.
```

Program 4.5 Cross Reference Generator.

```
 0 PROCEDURE search(x: INTEGER; VAR p: WPtr);
 1 BEGIN
 2    IF x < p↑.key THEN search(x, p↑.left)
 3    ELSIF x > p↑.key THEN search(x, p↑.right)
 4    ELSIF p # s THEN p↑.count := p↑.count + 1
 5    ELSE Allocate(p, SIZE(Word));
 6          WITH p↑ DO
 7             key := x; left := s; right := s; count := 1
 8          END
 9    END
10 END

Allocate            5
BEGIN               1
DO                  6
ELSE                5
ELSIF               3    4
END                 8    9    10
IF                  2
INTEGER             0
PROCEDURE           0
SIZE                5
THEN                2    3    4
VAR                 0
WITH                6
WPtr                0
Word                5
count               4    4    7
key                 2    3    7
left                2    7
p                   0    2    2    3    3    4    4
```

```
    4
                5    6
right           3    7
s               4    7    7
search          0    2    3
x               0    2    2    3    3    7
```

Table 4.4 Sample Output of Program 4.5.

4.4.4. Tree Deletion

We now turn to the inverse problem of insertion: *deletion.* Our task is to define an algorithm for deleting, i.e., removing the node with key x in a tree with ordered keys. Unfortunately, removal of an element is not generally as simple as insertion. It is straightforward if the element to be deleted is a terminal node or one with a single descendant. The difficulty lies in removing an element with two descendants, for we cannot point in two directions with a single pointer. In this situation, the deleted element is to be replaced by either the rightmost element of its left subtree or by the leftmost node of its right subtree, both of which have at most one descendant. The details are shown in the recursive procedure called *delete* (4.52). This procedure distinguishes among three cases:

1. There is no component with a key equal to x.
2. The component with key x has at most one descendant.
3. The component with key x has two descendants.

```
PROCEDURE delete(x: INTEGER; VAR p: Ptr);
  VAR q: Ptr:                                                    (4.52)

  PROCEDURE del (VAR r: Ptr);
  BEGIN
    IF r↑.right # NIL THEN del(r↑.right)
    ELSE q↑.key := r↑.key; q↑.count := r↑.count;
        q := r; r := r↑.left
    END
  END del;

BEGIN (*delete*)
  IF p = NIL THEN (*word is not in tree*)
  ELSIF x < p↑.key THEN delete(x, p↑.left)
  ELSIF x > p↑.key THEN delete(x, p↑.right)
  ELSE (*delete p↑*) q := p;
    IF q↑.right = NIL THEN p := q↑.left
    ELSIF q↑.left = NIL THEN p := q↑.right
    ELSE del(q↑.left)
    END ;
    (*Deallocate(q)*)
  END
END delete
```

The auxiliary, recursive procedure *del* is activated in case 3 only. It descends along the rightmost branch of the left subtree of the element q↑ to be deleted, and then it replaces the relevant information (key and count) in q↑ by the corresponding values of the rightmost component r↑ of that left subtree, whereafter t↑ may be disposed. The unspecified procedure *Deallocate* may be considered the inverse or opposite of *Allocate*. The latter allocates storage for a new component, but the former may be used to indicate to a computer system that storage occupied by q↑ is again free and available for other uses.

In order to illustrate the functioning of procedure (4.52), we refer to Fig. 4.28. Consider the tree (a); then delete successively the nodes with keys 13, 15, 5, 10. The resulting trees are shown in Fig. 4.28 (b-e).

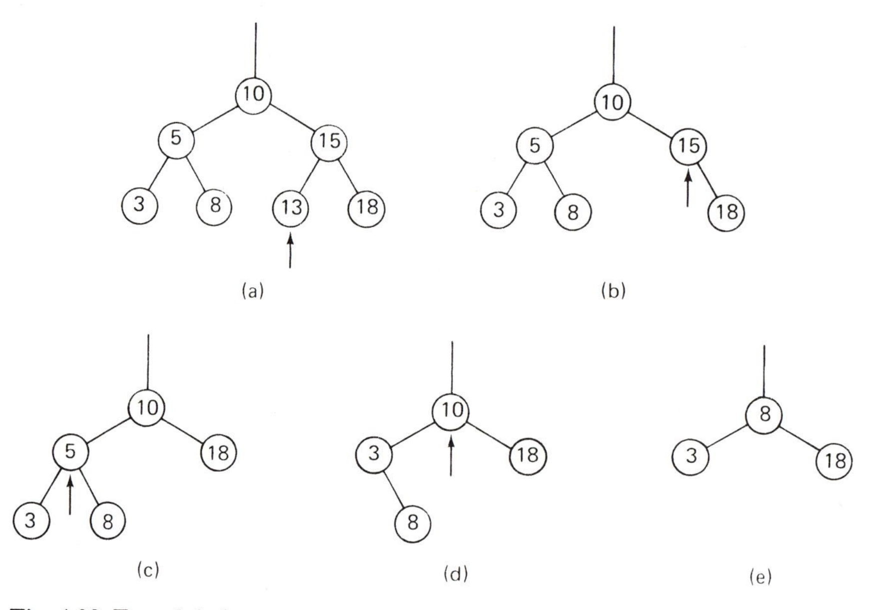

Fig. 4.28 Tree deletion.

4.4.5. Analysis of Tree Search and Insertion

It is a natural reaction to be suspicious of the algorithm of tree search and insertion. At least one should retain some skepticism until having been given a few more details about its behaviour. What worries many programmers at first is the peculiar fact that generally we do not know how the tree will grow; we have no idea about the shape that it will assume. We can only guess that it will most probably not be the perfectly balanced tree. Since the average number of comparisons needed to locate a key in a perfectly balanced tree with n nodes is approximately log n, the number of comparisons in a tree generated by this algorithm will be greater. But how much greater?

First of all, it is easy to find the worst case. Assume that all keys arrive in already strictly ascending (or descending) order. Then each key is appended immediately to the right (left) of its predecessor, and the resulting tree becomes completely degenerate, i.e., it turns out to be a linear list. The average search effort is then n/2 comparisons. This worst case evidently leads to a very poor performance of the search algorithm, and it seems to fully justify our skepticism. The remaining question is, of course, how likely this case will be. More precisely, we should like to know the length a_n of the search path averaged over all n keys and averaged over all n! trees that are generated from the n! permutations of the original n distinct keys. This problem of algorithmic analysis turns out to be fairly straightforward, and it is presented here as a typical example of analyzing an algorithm as well as for the practical importance of its result.

Given are n distinct keys with values 1, 2, ... , n. Assume that they arrive in a random order. The probability of the first key -- which notably becomes the root node -- having the value i is 1/n. Its left subtree will eventually contain i-1 nodes, and its right subtree n-i nodes (see Fig. 4.29). Let the average path length in the left subtree be denoted by a_{i-1}, and the one in the right subtree is a_{n-i}, again assuming that all possible permutations of the remaining n-1 keys are equally likely. The average path length in a tree with n nodes is the sum of the products of each node's level and its probability of access. If all nodes are assumed to be searched with equal likelihood, then

$$a_n = (Si: 1 \leq i \leq n : p_i) / n \qquad (4.53)$$

where p_i is the path length of node i.

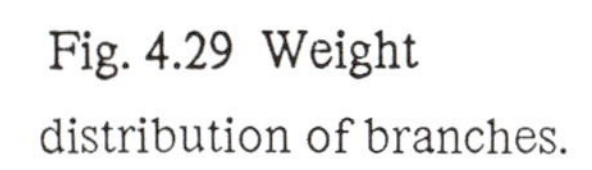

Fig. 4.29 Weight distribution of branches.

In the tree in Fig. 4.29 we divide the nodes into three classes:

1. The i-1 nodes in the left subtree have an average path length a_{i-1}
2. The root has a path length of 0.
3. The n-i nodes in the right subtree have an average path length a_{n-i}

Hence, (4.53) can be expressed as a sum of two terms 1) and 3)

$$a_n^{(i)} = ((i-1) * a_{i-1} + (n-i) * a_{n-i}) / n \qquad (4.54)$$

The desired quantity a_n is the average of $a_n^{(i)}$ over all i = 1 ... n, i.e., over all trees with the key 1, 2, ... , n at the root.

$$
\begin{aligned}
a_n &= (\mathrm{Si}: 1 \le i \le n : (i-1) * a_{i-1} + (n-i) * a_{n-i}) \,/\, n^2 \\
&= 2 * (\mathrm{Si}: 1 \le i \le n : (i-1) * a_{i-1}) \,/\, n^2 \\
&= 2 * (\mathrm{Si}: 1 \le i < n : i * a_i) \,/\, n^2
\end{aligned} \tag{4.55}
$$

Equation (4.55) is a recurrence relation of the form $a_n = f_1(a_1, a_2, \dots, a_n)$. From this we derive a simpler recurrence relation of the form $a_n = f_2(a_{n-1})$. From (4.55) we derive directly (1) by splitting off the last term, and (2) by substituting n-1 for n:

$$
\begin{aligned}
&(1)\ a_n = 2 * (n-1) * a_{n-1} / n^2 + 2 * (\mathrm{Si}: 1 \le i < n : i * a_i) / n^2 \\
&(2)\ a_{n-1} = 2 * (\mathrm{Si}: 1 \le i < n-1 : i * a_i) / (n-1)^2
\end{aligned}
$$

Multiplying (2) by $(n-1)^2/n^2$ yields

$$
(3)\ 2 * (\mathrm{Si}: 1 \le i < n-1 : i * a_i) / n^2 = a_{n-1} * (n-1)^2/n^2
$$

and substituting the right part of (3) in (1), we find

$$
\begin{aligned}
a_n &= 2 * (n-1) * a_{n-1} / n^2 + a_{n-1} * (n-1)^2/n^2 \\
&= a_{n-1} * (n-1)^2/n^2
\end{aligned} \tag{4.56}
$$

It turns out that a_n can be expressed in non-recursive, closed form in terms of the harmonic function

$$
\begin{aligned}
H_n &= 1 + 1/2 + 1/3 + \dots + 1/n \\
a_n &= 2 * (H_n*(n+1)/n - 1)
\end{aligned} \tag{4.57}
$$

From Euler's formula (using Euler's constant $g = 0.577\dots$)

$$
H_n = g + \ln n + 1/12n^2 + \dots
$$

we derive, for large n, the relationship

$$
a_n \doteq 2 * (\ln n + g - 1)
$$

Since the average path length in the perfectly balanced tree is approximately

$$
a_n' \doteq \log n - 1 \tag{4.58}
$$

we obtain, neglecting the constant terms which become insignificant for large n,

$$
\lim(a_n/a_n') = 2*\ln(n)/\log(n) = 2*\ln(2) \doteq 1.386\dots \tag{4.59}
$$

What does the result (4.59) of this analysis teach us? It tells us that by taking the pains of always constructing a perfectly balanced tree instead of the random tree obtained from Program 4.4, we could -- always provided that all keys are looked up with equal probability -- expect an average improvement in the search path length of at most 39%. Emphasis is to be put on the word average, for the improvement may of course be very much greater in the unhappy case in which the generated tree had completely degenerated into a list, which, however, is very unlikely to occur. In this connection it is noteworthy that the expected average path length of the random tree grows also strictly logarithmically with the number of its nodes, even though the worst case path length grows linearly.

The figure of 39% imposes a limit on the amount of additional effort that may be spent profitably on any kind of reorganization of the tree's structure upon insertion of keys. Naturally, the ratio between the frequencies of access (retrieval) of nodes (information) and of insertion (update) significantly influences the payoff limits of any such undertaking. The higher this ratio, the higher is the payoff of a reorganization procedure. The 39% figure is low enough that in most applications improvements of the straight tree insertion algorithm do not pay off unless the number of nodes and the access vs. insertion ratio are large.

4.5. BALANCED TREES

From the preceding discussion it is clear that an insertion procedure that always restores the trees' structure to perfect balance has hardly any chance of being profitable, because the restoration of perfect balance after a random insertion is a fairly intricate operation. Possible improvements lie in the formulation of less strict definitions of balance. Such imperfect balance criteria should lead to simpler tree reorganization procedures at the cost of only a slight deterioration of average search performance. One such definition of balance has been postulated by Adelson-Velskii and Landis [4-1]. The balance criterion is the following:

A tree is *balanced* if and only if for every node the heights of its two subtrees differ by at most 1.

Trees satisfying this condition are often called AVL-trees (after their inventors). We shall simply call them *balanced trees* because this balance criterion appears a most suitable one. (Note that all perfectly balanced trees are also AVL-balanced.)

The definition is not only simple, but it also leads to a manageable rebalancing procedure and an average search path length practically identical to that of the perfectly balanced tree. The following operations can be performed on balanced trees in O(log n) units of time, even in the worst case:

1. Locate a node with a given key.
2. Insert a node with a given key.
3. Delete the node with a given key.

These statements are direct consequences of a theorem proved by Adelson-Velskii and Landis, which guarantees that a balanced tree will never be more than 45% higher than its perfectly balanced counterpart, no matter how many nodes there are. If we denote the height of a balanced tree with n nodes by $h_b(n)$, then

$$\log(n+1) \leq h_b(n) < 1.4404*\log(n+2) - 0.328 \qquad (4.60)$$

The optimum is of course reached if the tree is perfectly balanced for $n = 2^k-1$. But which is the structure of the *worst* AVL-balanced tree? In order to find the maximum height h of all balanced trees with n nodes, let us consider a fixed height h and try to construct the balanced tree with the minimum number of nodes. This strategy is recommended because, as in the case of the minimal height, the value can be attained only for certain specific values of n. Let this tree of height h be denoted by T_h. Clearly, T_0 is the empty tree, and T_1 is the tree with a single node. In order to construct the tree T_h for $h > 1$, we will provide the root with two subtrees which again have a minimal number of nodes. Hence, the subtrees are also T's. Evidently, one subtree must have height h-1, and the other is then allowed to have a height of one less, i.e. h-2. Figure 4.30 shows the trees with height 2, 3, and 4. Since their composition principle very strongly resembles that of Fibonacci numbers, they are called *Fibonacci-trees*. They are defined as follows:

1. The empty tree is the Fibonacci-tree of height 0.

2. A single node is the Fibonacci-tree of height 1.
3. If T_{h-1} and T_{h-2} are Fibonacci-trees of heights h-1 and h-2, then
   $T_h = \langle T_{h-1}, x, T_{h-2} \rangle$ is a Fibonacci-tree.
4. No other trees are Fibonacci-trees.

The number of nodes of T_h is defined by the following simple recurrence relation:

$$N_0 = 0, \; N_1 = 1$$
$$N_h = N_{h-1} + 1 + N_{h-2} \qquad (4.61)$$

The N_i are those numbers of nodes for which the worst case (upper limit of h) of (4.60) can be attained, and they are called *Leonardo numbers*.

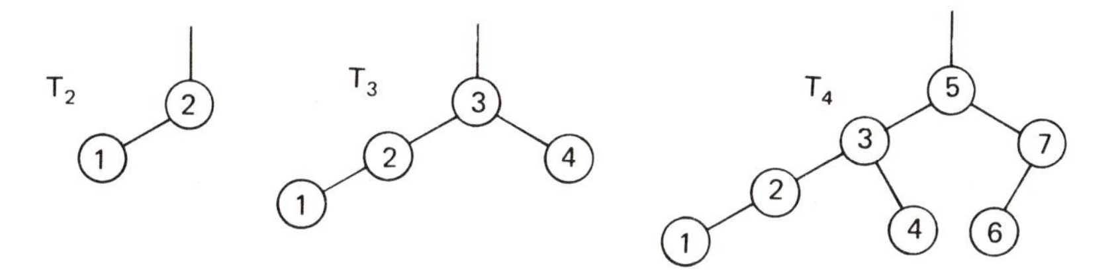

Fig. 4.30 Fibonacci-trees of height 2, 3, and 4.

4.5.1. Balanced Tree Insertion

Let us now consider what may happen when a new node is inserted in a balanced tree. Given a root r with the left and right subtrees L and R, three cases must be distinguished. Assume that the new node is inserted in L causing its height to increase by 1:

1. $h_L = h_R$: L and R become of unequal height, but the balance criterion is not violated.
2. $h_L < h_R$: L and R obtain equal height, i.e., the balance has even been improved.
3. $h_L > h_R$: the balance criterion is violated, and the tree must be restructured.

Consider the tree in Fig. 4.31. Nodes with keys 9 and 11 may be inserted without rebalancing; the tree with root 10 will become one-sided (case 1); the one with root 8 will improve its balance (case 2). Insertion of nodes 1, 3, 5, or 7, however, requires subsequent rebalancing.

Some careful scrutiny of the situation reveals that there are only two essentially different constellations needing individual treatment. The remaining ones can be derived by symmetry considerations from those two. Case 1 is characterized by inserting keys 1 or 3 in the tree of Fig. 4.31, case 2 by inserting nodes 5 or 7.

The two cases are generalized in Fig. 4.32 in which rectangular boxes denote subtrees, and the height added by the insertion is indicated by crosses. Simple transformations of the two structures restore the desired balance. Their result is shown in Fig. 4.33; note that the

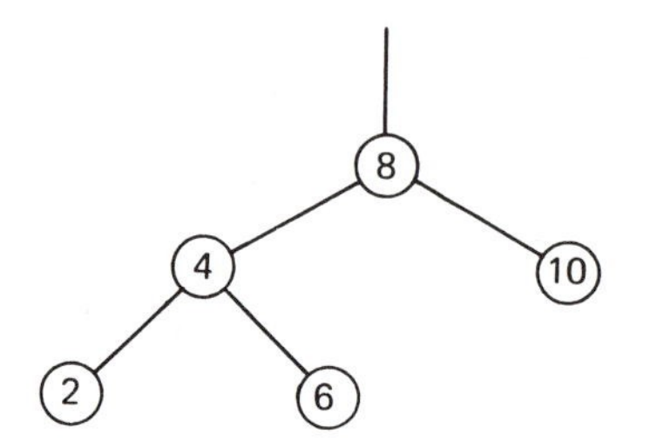

Fig. 4.31 Balanced tree.

only movements allowed are those occurring in the vertical direction, whereas the relative horizontal positions of the shown nodes and subtrees must remain unchanged.

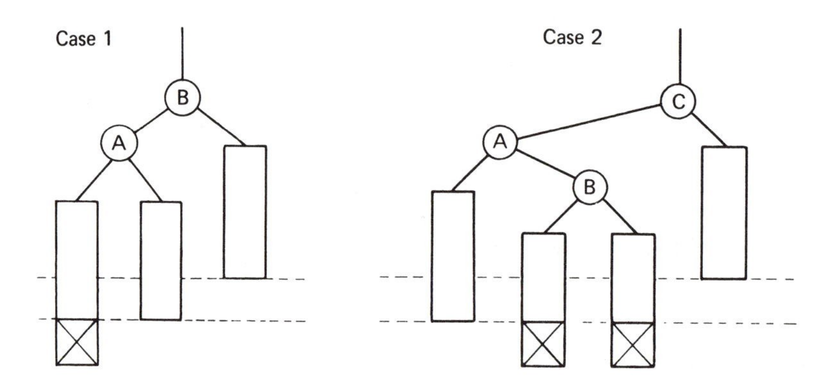

Fig. 4.32 Imbalance resulting from insertion.

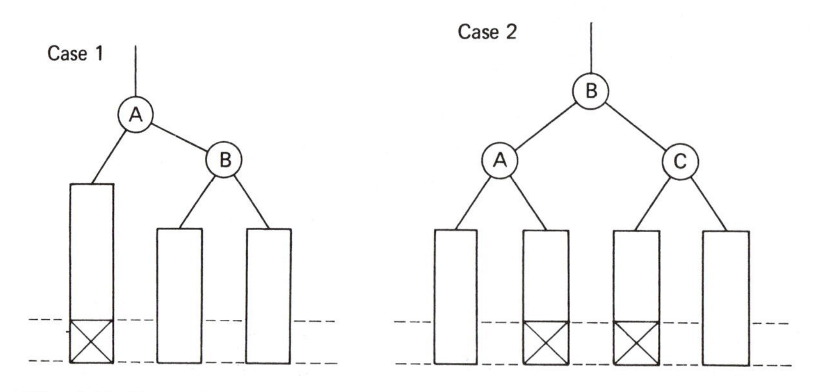

Fig. 4.33 Restoring the balance.

An algorithm for insertion and rebalancing critically depends on the way information about the tree's balance is stored. An extreme solution lies in keeping balance information entirely implicit in the tree structure itself. In this case, however, a node's balance factor must be rediscovered each time it is affected by an insertion, resulting in an excessively high overhead. The other extreme is to attribute an explicitly stored balance factor to every node. The definition (4.48) of the type *Node* is then extended into

```
TYPE Ptr =      POINTER TO Node;
TYPE Balance =[-1 .. +1];
TYPE Node =     RECORD key: INTEGER;                              (4.62)
                  count: INTEGER;
                  left, right: Ptr
                  bal: Balance
                END
```

We shall subsequently interpret a node's balance factor as the height of its right subtree minus the height of its left subtree, and we shall base the resulting algorithm on the node type (4.62). The process of node insertion consists essentially of the following three consecutive parts:

1. Follow the search path until it is verified that the key is not already in the tree.
2. Insert the new node and determine the resulting balance factor.
3. Retreat along the search path and check the balance factor at each node. Rebalance if necessary.

Although this method involves some redundant checking (once balance is established, it need not be checked on that node's ancestors), we shall first adhere to this evidently correct schema because it can be implemented through a pure extension of the already established search and insertion procedure of Program 4.4. This procedure describes the search operation needed at each single node, and because of its recursive formulation it can easily accommodate an additional operation on the way back along the search path. At each step, information must be passed as to whether or not the height of the subtree (in which the insertion had been performed) had increased. We therefore extend the procedure's parameter list by the Boolean h with the meaning *the subtree height has increased*. Clearly, h must denote a variable parameter since it is used to transmit a result.

Assume now that the process is returning to a node p↑ from the left branch (see Fig. 4.32), with the indication that it has increased its height. We now must distinguish between the three conditions involving the subtree heights prior to insertion:

1. $h_L < h_R$, p↑.bal = +1, the previous imbalance at p has been equilibrated.
2. $h_L = h_R$, p↑.bal = 0, the weight is now slanted to the left.
3. $h_L > h_R$, p↑.bal = -1, rebalancing is necessary.

In the third case, inspection of the balance factor of the root of the left subtree (say, p1↑.bal) determines whether case 1 or case 2 of Fig. 4.32 is present. If that node has also a higher left than right subtree, then we have to deal with case 1, otherwise with case 2. (Convince yourself that a left subtree with a balance factor equal to 0 at its root cannot occur in this case.) The rebalancing operations necessary are entirely expressed as sequences of pointer reassignments. In fact, pointers are cyclically exchanged, resulting in either a single or a double rotation of the two or three nodes involved. In addition to pointer rotation, the respective node balance factors have to be updated. The details are shown in the search, insertion, and rebalancing procedure (4.63).

The working principle is shown by Fig. 4.34. Consider the binary tree (a) which consists of two nodes only. Insertion of key 7 first results in an unbalanced tree (i.e., a linear list). Its balancing involves a RR single rotation, resulting in the perfectly balanced tree (b). Further insertion of nodes 2 and 1 result in an imbalance of the subtree with root 4. This subtree is balanced by an LL single rotation (d). The subsequent insertion of key 3 immediately offsets the balance criterion at the root node 5. Balance is thereafter reestablished by the more complicated LR double rotation; the outcome is tree (e). The only candidate for losing balance after a next insertion is node 5. Indeed, insertion of node 6 must invoke the fourth case of rebalancing outlined in (4.63), the RL double rotation. The final tree is shown in Fig.4.34 (f).

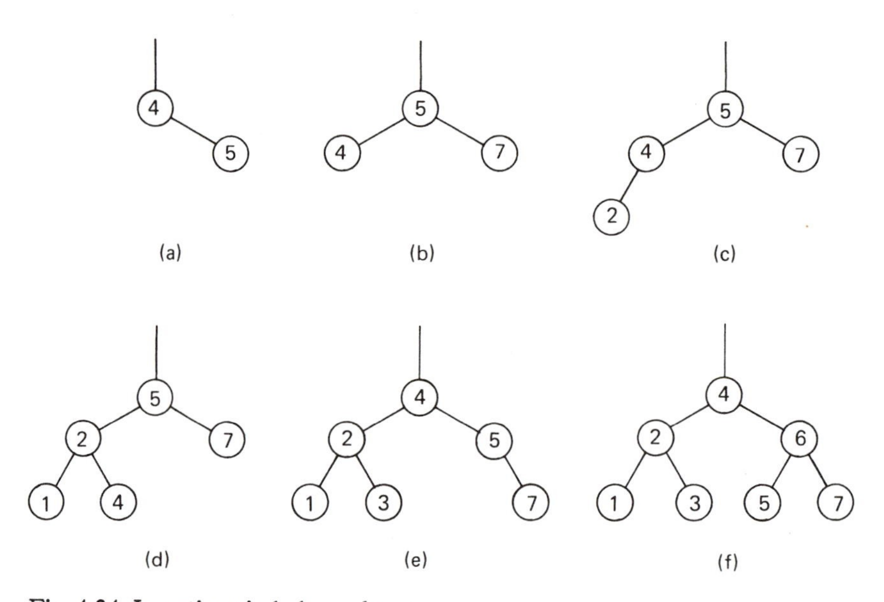

Fig. 4.34 Insertions in balanced tree.

```
PROCEDURE search(x: INTEGER; VAR p: Ptr; VAR h: BOOLEAN);
  VAR p1, p2: Ptr;  (*~h*)
BEGIN                                                          (4.63)
  IF p = NIL THEN (*insert*)
    ALLOCATE(p, SIZE(Node)); h := TRUE;
    WITH p↑ DO
      key := x; count := 1; left := NIL; right := NIL; bal := 0
    END
  ELSIF p↑.key > x THEN
    search(x, p↑.left, h);
    IF h THEN  (*left branch has grown*)
```

```
      CASE p↑.bal OF
       1: p↑.bal := 0; h := FALSE |
       0: p↑.bal := -1 |
      -1: (*rebalance*) p1 := p↑.left;
         IF p1↑.bal = -1 THEN (*single LL rotation*)
          p↑.left := p1↑.right; p1↑.right := p;
          p↑.bal := 0; p := p1
         ELSE (*double LR rotation*) p2 := p1↑.right;
          p1↑.right := p2↑.left; p2↑.left := p1;
          p↑.left := p2↑.right; p2↑.right := p;
          IF p2↑.bal = -1 THEN p↑.bal := 1 ELSE p↑.bal := 0 END ;
          IF p2↑.bal = +1 THEN p1↑.bal := -1 ELSE p1↑.bal := 0 END ;
          p := p2
         END ;
         p↑.bal := 0; h := FALSE
      END
     END
    ELSIF p↑.key < x THEN
     search(x, p↑.right, h);
     IF h THEN (*right branch has grown*)
      CASE p↑.bal OF
      -1: p↑.bal := 0; h := FALSE |
       0: p↑.bal := 1 |
       1: (*rebalance*) p1 := p↑.right;
         IF p1↑.bal = 1 THEN (*single RR rotation*)
          p↑.right := p1↑.left; p1↑.left := p;
          p↑.bal := 0; p := p1
         ELSE (*double RL rotation*) p2 := p1↑.left;
          p1↑.left := p2↑.right; p2↑.right := p1;
          p↑.right := p2↑.left; p2↑.left := p;
          IF p2↑.bal = +1 THEN p↑.bal := -1 ELSE p↑.bal := 0 END ;
          IF p2↑.bal = -1 THEN p1↑.bal := 1 ELSE p1↑.bal := 0 END ;
          p := p2
         END ;
         p↑.bal := 0; h := FALSE
      END
     END
    ELSE p↑.count := p↑.count + 1
    END
  END search
```

Two particularly interesting questions concerning the performance of the balanced tree insertion algorithm are the following:

1. If all n! permutations of n keys occur with equal probability, what is the expected height of the constructed balanced tree?

2. What is the probability that an insertion requires rebalancing?

Mathematical analysis of this complicated algorithm is still an open problem. Empirical tests support the conjecture that the expected height of the balanced tree generated by (4.63) is $h = \log(n)+c$, where c is a small constant ($c \doteq 0.25$). This means that in practice the AVL-balanced tree behaves as well as the perfectly balanced tree, although it is much simpler to maintain. Empirical evidence also suggests that, on the average, rebalancing is necessary once for approximately every two insertions. Here single and double rotations are equally probable. The example of Fig. 4.34 has evidently been carefully chosen to demonstrate as many rotations as possible in a minimum number of insertions.

The complexity of the balancing operations suggests that balanced trees should be used only if information retrievals are considerably more frequent than insertions. This is particularly true because the nodes of such search trees are usually implemented as densely packed records in order to economize storage. The speed of access and of updating the balance factors -- each requiring two bits only -- is therefore often a decisive factor to the efficiency of the rebalancing operation. Empirical evaluations show that balanced trees lose much of their appeal if tight record packing is mandatory. It is indeed difficult to beat the straightforward, simple tree insertion algorithm.

4.5.2. Balanced Tree Deletion

Our experience with tree deletion suggests that in the case of balanced trees deletion will also be more complicated than insertion. This is indeed true, although the rebalancing operation remains essentially the same as for insertion. In particular, rebalancing consists again of either single or a double rotations of nodes.

The basis for balanced tree deletion is algorithm (4.52). The easy cases are terminal nodes and nodes with only a single descendant. If the node to be deleted has two subtrees, we will again replace it by the rightmost node of its left subtree. As in the case of insertion (4.63), a Boolean variable parameter h is added with the meaning *the height of the subtree has been reduced*. Rebalancing has to be considered only when h is true. h is made true upon finding and deleting a node, or if rebalancing itself reduces the height of a subtree. In (4.64) we introduce the two (symmetric) balancing operations in the form of procedures, because they have to be invoked from more than one point in the deletion algorithm. Note that *balanceL* is applied when the left, *balanceR* after the right branch had been reduced in height.

The operation of the procedure is illustrated in Fig. 4.35. Given the balanced tree (a), successive deletion of the nodes with keys 4, 8, 6, 5, 2, 1, and 7 results in the trees (b) ... (h). Deletion of key 4 is simple in itself, because it represents a terminal node. However, it results in an unbalanced node 3. Its rebalancing operation invoves an LL single rotation. Rebalancing becomes again necessary after the deletion of node 6. This time the right subtree of the root (7) is rebalanced by an RR single rotation. Deletion of node 2, although in itself straightforward since it has only a single descendant, calls for a complicated RL double rotation. The fourth case, an LR double rotation, is finally invoked after the removal of node 7, which at first was replaced by the rightmost element of its left subtree, i.e., by the node with key 3.

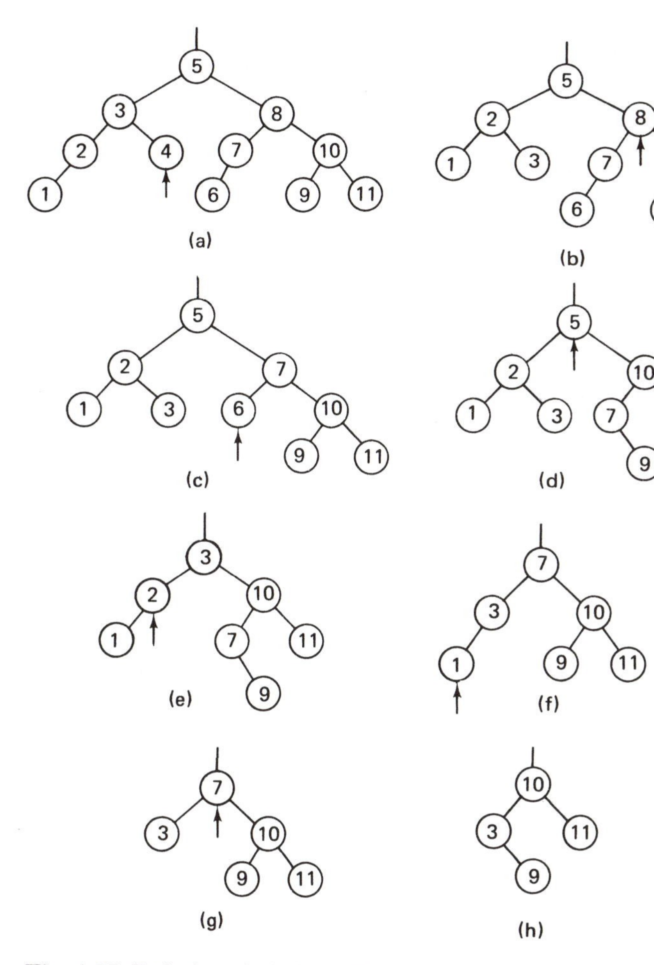

Fig. 4.35 Deletions in balanced tree.

```
PROCEDURE balanceL(VAR p: Ptr; VAR h: BOOLEAN);
 VAR p1, p2: Ptr; b1, b2: Balance;                          (4.64)
BEGIN (*h; left branch has shrunk*)
 CASE p↑.bal OF
  -1: p↑.bal := 0 |
   0: p↑.bal := 1; h := FALSE |
   1: (*rebalance*) p1 := p↑.right; b1 := p1↑.bal;
     IF b1 >= 0 THEN  (*single RR rotation*)
       p↑.right := p1↑.left; p1↑.left := p;
       IF b1 = 0 THEN p↑.bal := 1; p1↑.bal := -1; h := FALSE
         ELSE p↑.bal := 0; p1↑.bal := 0
       END ;
       p := p1
```

```
    ELSE (*double RL rotation*)
      p2 := p1↑.left; b2 := p2↑.bal;
      p1↑.left := p2↑.right; p2↑.right := p1;
      p↑.right := p2↑.left; p2↑.left := p;
      IF b2 = +1 THEN p↑.bal := -1 ELSE p↑.bal := 0 END ;
      IF b2 = -1 THEN p1↑.bal := 1 ELSE p1↑.bal := 0 END ;
      p := p2; p2↑.bal := 0
    END
  END
END balanceL;

PROCEDURE balanceR(VAR p: Ptr; VAR h: BOOLEAN);
  VAR p1, p2: Ptr; b1, b2: Balance;
BEGIN (*h; right branch has shrunk*)
  CASE p↑.bal OF
   1: p↑.bal := 0 |
   0: p↑.bal := -1; h := FALSE |
  -1: (*rebalance*) p1 := p↑.left; b1 := p1↑.bal;
    IF b1 <= 0 THEN (*single LL rotation*)
      p↑.left := p1↑.right; p1↑.right := p;
      IF b1 = 0 THEN p↑.bal := -1; p1↑.bal := 1; h := FALSE
        ELSE p↑.bal := 0; p1↑.bal := 0
      END ;
      p := p1
    ELSE (*double LR rotation*)
      p2 := p1↑.right; b2 := p2↑.bal;
      p1↑.right := p2↑.left; p2↑.left := p1;
      p↑.left := p2↑.right; p2↑.right := p;
      IF b2 = -1 THEN p↑.bal := 1 ELSE p↑.bal := 0 END ;
      IF b2 = +1 THEN p1↑.bal := -1 ELSE p1↑.bal := 0 END ;
      p := p2; p2↑.bal := 0
    END
  END
END balanceR;

PROCEDURE delete(x: INTEGER; VAR p: Ptr; VAR h: BOOLEAN);
  VAR q: Ptr;

  PROCEDURE del(VAR r: Ptr; VAR h: BOOLEAN);
  BEGIN (*~h*)
    IF r↑.right # NIL THEN
      del(r↑.right, h);
      IF h THEN balanceR(r, h) END
    ELSE q↑.key := r↑.key; q↑.count := r↑.count;
      q := r; r := r↑.left; h := TRUE
    END
  END del;
```

```
BEGIN (*~h*)
 IF p = NIL THEN (*key not in tree*)
 ELSIF p↑.key > x THEN
  delete(x, p↑.left, h);
  IF h THEN balanceL(p, h) END
 ELSIF p↑.key < x THEN
  delete(x, p↑.right, h);
  IF h THEN balanceR(p, h) END
 ELSE (*delete p↑*) q := p;
  IF q↑.right = NIL THEN p := q↑.left; h := TRUE
  ELSIF q↑.left = NIL THEN p := q↑.right; h := TRUE
  ELSE del(q↑.left, h);
   IF h THEN balanceL(p, h) END
  END;
  (* Deallocate(q) *)
 END
END delete
```

Fortunately, deletion of an element in a balanced tree can also be performed with -- in the worst case -- O(log n) operations. An essential difference between the behaviour of the insertion and deletion procedures must not be overlooked, however. Whereas insertion of a single key may result in at most one rotation (of two or three nodes), deletion may require a rotation at every node along the search path. Consider, for instance, deletion of the rightmost node of a Fibonacci-tree. In this case the deletion of any single node leads to a reduction of the height of the tree; in addition, deletion of its rightmost node requires the maximum number of rotations. This therefore represents the worst choice of node in the worst case of a balanced tree, a rather unlucky combination of chances. How probable are rotations, then, in general?

The surprising result of empirical tests is that whereas one rotation is invoked for approximately every two insertions, one is required for every five deletions only. Deletion in balanced trees is therefore about as easy -- or as complicated -- as insertion.

4.6. OPTIMAL SEARCH TREES

So far our consideration of organizing search trees has been based on the assumption that the frequency of access is equal for all nodes, that is, that all keys are equally probable to occur as a search argument. This is probably the best assumption if one has no idea of access distribution. However, there are cases (they are the exception rather than the rule) in which information about the probabilities of access to individual keys is available. These cases usually have the characteristic that the keys always remain the same, i.e., the search tree is subjected neither to insertion nor deletion, but retains a constant structure. A typical example is the scanner of a compiler which determines for each word (identifier) whether or not it is a keyword (reserved word). Statistical measurements over hundreds of compiled programs may in this case yield accurate information on the relative frequencies of occurrence, and thereby of access, of individual keys.

Assume that in a search tree the probability with which node i is accessed is

$$Pr\{x = k_i\} = p_i, \qquad (Si: 1 \leq i \leq n : p_i) = 1 \qquad (4.65)$$

We now wish to organize the search tree in a way that the total number of search steps -- counted over sufficiently many trials -- becomes minimal. For this purpose the definition of path length (4.34) is modified by (1) attributing a certain weight to each node and by (2) assuming the root to be at level 1 (instead of 0), because it accounts for the first comparison along the search path. Nodes that are frequently accessed become heavy nodes; those that are rarely visited become light nodes. The (internal) *weighted path length* is then the sum of all paths from the root to each node weighted by that node's probability of access.

$$P = Si: 1 \leq i \leq n : p_i*h_i \qquad (4.66)$$

h_i is the level of node i. The goal is now to minimize the weighted path length for a given probability distribution. As an example, consider the set of keys 1, 2, 3, with probabilities of access $p_1 = 1/7$, $p_2 = 2/7$, and $p_3 = 4/7$. These three keys can be arranged in five different ways as search trees (see Fig. 4.36).

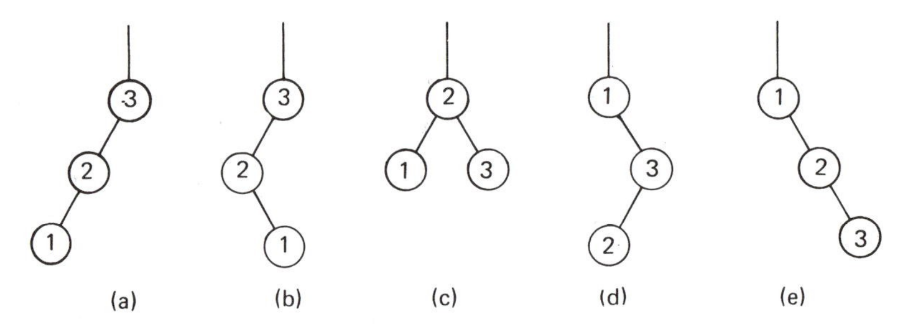

Fig. 4.36 Search trees with three nodes.

The weighted path lengths of trees (a) to (e) are computed according to (4.66) as

$$P(a) = 11/7,\ P(b) = 12/7,\ P(c) = 12/7,\ P(d) = 15/7,\ P(e) = 17/7$$

Hence, in this example, not the perfectly balanced tree (c), but the degenerate tree (a) turns out to be optimal.

The example of the compiler scanner immediately suggests that this problem should be viewed under a slightly more general condition: words occurring in the source text are not always keywords; as a matter of fact, their being keywords is rather the exception. Finding that a given word k is not a key in the search tree can be considered as an access to a hypothetical "special node" inserted between the next lower and next higher key (see Fig. 4.19) with an associated external path length. If the probability q_i of a search argument x lying between the two keys k_i and k_{i+1} is also known, this information may considerably change the structure of the optimal search tree. Hence, we generalize the problem by also considering unsuccessful searches. The overall average weighted path length is now

$$P = (Si: 1 \leq i \leq n : p_i * h_i) + (Sj: 0 \leq j \leq m : q_j * h'_j) \qquad (4.67)$$

where

$$(Si: 1 \leq i \leq n : p_i) + (Sj: 0 \leq j \leq m : q_j) = 1$$

and where, h_i is the level of the (internal) node i and h'_j is the level of the external node j. The average weighted path length may be called the *cost* of the search tree, since it represents a measure for the expected amount of effort to be spent for searching. The search tree that requires the minimal cost among all trees with a given set of keys k_i and probabilities p_i and q_j is called the *optimal tree.*

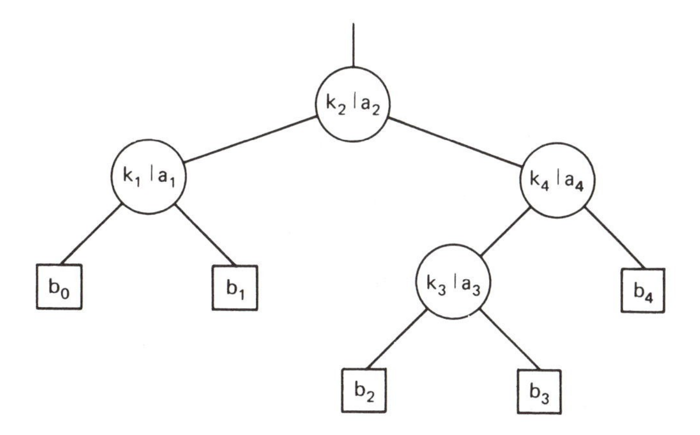

Fig. 4.37 Search tree with associated access frequencies.

For finding the optimal tree, there is no need to require that the p's and q's sum up to 1. In fact, these probabilities are commonly determined by experiments in which the accesses to nodes are counted. Instead of using the probabilities p_i and q_j, we will subsequently use such frequency counts and denote them by

a_i = number of times the search argument x equals k_i

b_j = number of times the search argument x lies between k_j and k_{j+1}

By convention, b_0 is the number of times that x is less than k_1, and b_n is the frequency of x being greater than k_n (see Fig. 4.37). We will subsequently use P to denote the *accumulated weighted path length* instead of the average path length:

$$P = (Si: 1 \le i \le n : a_i*h_i) + (Sj: 0 \le j \le m : b_j*h'_j) \qquad (4.68)$$

Thus, apart from avoiding the computation of the probabilities from measured frequency counts, we gain the further advantage of being able to use integers instead of fractions in our search for the optimal tree.

Considering the fact that the number of possible configurations of n nodes grows exponentially with n, the task of finding the optimum seems rather hopeless for large n. Optimal trees, however, have one significant property that helps to find them: all their subtrees are optimal too. For instance, if the tree in Fig. 4.37 is optimal, then the subtree with keys k_3 and k_4 is also optimal as shown. This property suggests an algorithm that systematically finds larger and larger trees, starting with individual nodes as smallest possible subtrees. The tree thus grows from the leaves to the root, which is, since we are used to drawing trees upside-down, the bottom-up direction [4-6].

The equation that is the key to this algorithm is (4.69). Let P be the weighted path length of a tree, and let P_L and P_R be those of the left and right subtrees of its root. Clearly, P is the sum of P_L and P_R, and the number of times a search travels on the leg to the root, which is simply the total number W of search trials. We call W the *weight* of the tree. Its average path length is then P/W.

$$P = P_L + W + P_R \qquad (4.69)$$

$$W = (Si: 1 \le i \le n : a_i) + (Sj: 0 \le j \le m : b_j) \qquad (4.70)$$

These considerations show the need for a denotation of the weights and the path lengths of any subtree consisting of a number of adjacent keys. Let T_{ij} be the optimal subtree consisting of nodes with keys $k_{i+1}, k_{i+2}, \dots, k_j$. Then let w_{ij} denote the weight and let p_{ij} denote the path length of T_{ij}. Clearly $P = p_{0,n}$ and $W = w_{0,n}$. These quantities are defined by the recurrence relations (4.71) and (4.72)

$$\begin{aligned} w_{ii} &= b_i && (0 \le i \le n) \\ w_{ij} &= w_{i,j-1} + a_j + b_j && (0 \le i < j \le n) \end{aligned} \qquad (4.71)$$

$$\begin{aligned} p_{ii} &= w_{ii} && (0 \le i \le n) \\ p_{ij} &= w_{ij} + \textbf{MIN}\ k: i < k \le j : (p_{i,k-1} + p_{kj}) && (0 \le i < j \le n) \end{aligned} \qquad (4.72)$$

The last equation follows immediately from (4.69) and the definition of optimality. Since there are approximately $n^2/2$ values p_{ij}, and because (4.72) calls for a choice among all cases such that $0 < j\text{-}i \leq n$, the minimization operation will involve approximately $n^3/6$ operations. Knuth pointed out that a factor n can be saved by the following consideration, which alone makes this algorithm usable for practical purposes.

Let r_{ij} be a value of k which achieves the minimum in (4.72). It is possible to limit the search for r_{ij} to a much smaller interval, i.e., to reduce the number of the j-i evaluation steps. The key is the observation that if we have found the root r_{ij} of the optimal subtree T_{ij}, then neither extending the tree by adding a node at the right, nor shrinking the tree by removing its leftmost node ever can cause the optimal root to move to the left. This is expressed by the relation

$$r_{i,j-1} \leq r_{ij} \leq r_{i+1,j} \tag{4.73}$$

which limits the search for possible solutions for r_{ij} to the range $r_{i,j-1} \dots r_{i+1,j}$. This results in a total number of elementary steps in the order of n^2.

We are now ready to construct the optimization algorithm in detail. We recall the following definitions, which are based on optimal trees T_{ij} consisting of nodes with keys $k_{i+1} \dots k_j$.

1. a_i: the frequency of a search for k_i.
2. b_j: the frequency of a search argument x between k_j and k_{j+1}.
3. w_{ij}: the weight of T_{ij}.
4. p_{ij}: the weighted path length of T_{ij}.
5. r_{ij}: the index of the root of T_{ij}.

Given

```
TYPE index = [0 .. n]
```

we declare the following arrays:

```
a:    ARRAY [1 .. n] OF CARDINAL;
b:    ARRAY index OF CARDINAL;
p,w:  ARRAY index, index OF CARDINAL;                    (4.74)
r:    ARRAY index, index OF index
```

Assume that the weights w_{ij} have been computed from a and b in a straightforward way [see (4.71)]. Now consider w as the argument of the procedure *OptTree* to be developed and consider r as its result, because r describes the tree structure completely. p may be considered an intermediate result. Starting out by considering the smallest possible subtrees, namely those consisting of no nodes at all, we proceed to larger and larger trees. Let us denote the width j-i of the subtree T_{ij} by h. Then we can trivially determine the values p_{ii} for all trees with h = 0 according to (4.72).

```
FOR i := 0 TO n DO p[i,i] := b[i] END                    (4.75)
```

In the case h = 1 we deal with trees consisting of a single node, which plainly is also the root (see Fig. 4.38).

```
FOR i := 0 TO n-1 DO
  j := i+1; p[i,j] := w[i,j] + p[i,i] + p[j,j]; r[i,j] := j        (4.76)
END
```

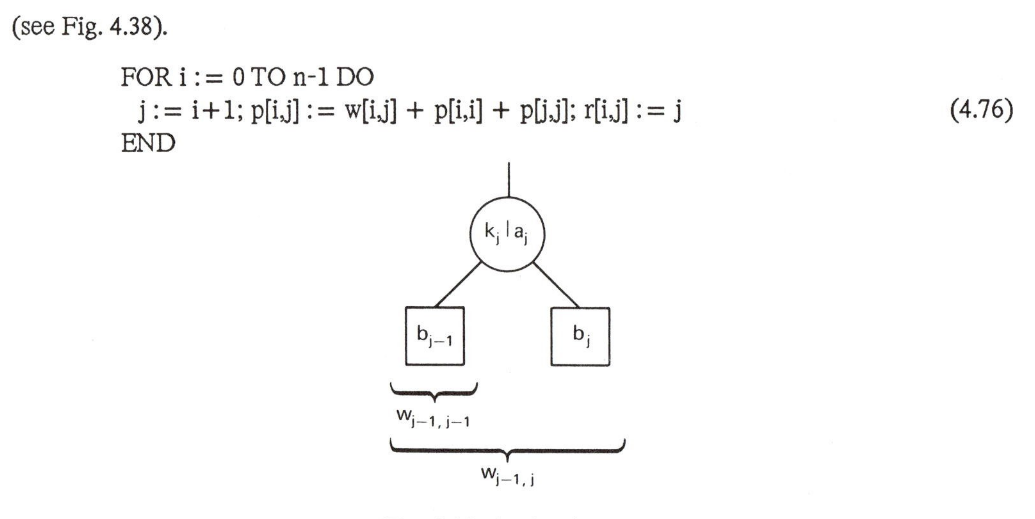

Fig. 4.38 Optimal tree
with one node.

Note that i denotes the left index limit and j the right index limit in the considered tree T_{ij}. For the cases $h > 1$ we use a repetitive statement with h ranging from 2 to n, the case $h = n$ spanning the entire tree $T_{0,n}$. In each case the minimal path length p_{ij} and the associated root index r_{ij} are determined by a simple repetitive statement with an index k ranging over the interval given by (4.73).

```
FOR h := 2 TO n DO
  FOR i := 0 TO n-h DO                                              (4.77)
    j := i+h;
    find k and min = MIN k: i < k ≤ j : (p[i,k-1] + p[k,j])
    such that r[i,j-1] ≤ k ≤ r[i+1,j];
    p[i,j] := min + w[i,j]; r[i,j] := k
  END
END
```

The details of the refinement of the statement in italics can be found in Program 4.6. The average path length of $T_{0,n}$ is now given by the quotient $p_{0,n}/w_{0,n}$, and its root is the node with index $r_{0,n}$.

Let us now describe the structure of Program 4.6. Its two main components are the procedures to find the optimal search tree, given a weight distribution w, and to display the tree given the indices r. First, the counts a and b and the keys are read from an input source. The keys are actually not involved in the computation of the tree structure; they are merely used in the subsequent display of the tree. After printing the frequency statistics, the program proceeds to compute the path length of the perfectly balanced tree, in passing also determining the roots of its subtrees. Thereafter, the average weighted path length is printed and the tree is displayed.

In the third part, procedure *OptTree* is activated in order to compute the optimal search tree; thereafter, the tree is displayed. And finally, the same procedures are used to compute and display the optimal tree considering the key frequencies only, ignoring the frequencies of non-keys.

Table 4.5 and Figs. 4.40 through 4.42 show the results generated by Program 4.6 when applied to its own program text. The differences in the three figures demonstrate that the balanced tree cannot even be considered as nearly optimal, and that the frequencies of the non-keys crucially influence the choice of the optimal structure.

```
MODULE OptTree;
 FROM InOut IMPORT
  OpenInput, OpenOutput, Read, ReadCard, ReadString, WriteCard,
  WriteString, Write, WriteLn, Done, CloseInput, CloseOutput;
 FROM Storage IMPORT ALLOCATE;

 CONST N = 100; (*max no. of keywords*)
     WL = 16; (*max keyword length*)

 TYPE Word = ARRAY [0 .. WL-1] OF CHAR;
    index = [0 .. N];

 VAR ch: CHAR;
  i, j, n: CARDINAL;
  key: ARRAY index OF Word;
  a:   ARRAY index OF CARDINAL;
  b:   ARRAY index OF CARDINAL;
  p,w: ARRAY index, index OF CARDINAL;
  r:   ARRAY index, index OF CARDINAL;

PROCEDURE BalTree(i,j: CARDINAL): CARDINAL;
 VAR k: CARDINAL;
BEGIN k := (i+j+1) DIV 2; r[i,j] := k;
 IF i >= j THEN RETURN 0
   ELSE RETURN BalTree(i,k-1) + BalTree(k,j) + w[i,j]
 END
END BalTree;

PROCEDURE OptTree;
 VAR x, min: CARDINAL;
  i, j, k, h, m: CARDINAL;
BEGIN (*argument: W, results: p, r*)
 FOR i := 0 TO n DO p[i,i] := 0 END ;
 FOR i := 0 TO n-1 DO
  j := i+1; p[i,j] := w[i,j]; r[i,j] := j
 END ;
 FOR h := 2 TO n DO
  FOR i := 0 TO n-h DO
```

```
      j := i+h; m := r[i,j-1]; min := p[i,m-1] + p[m,j];
      FOR k := m+1 TO r[i+1,j] DO
        x := p[i,k-1] + p[k,j];
        IF x < min THEN
          m := k; min := x
        END
      END ;
      p[i,j] := min + w[i,j]; r[i,j] := m
    END
  END
END OptTree;

PROCEDURE PrintTree(i, j, level: CARDINAL);
  VAR k: CARDINAL;
BEGIN
  IF i < j THEN
    PrintTree(i, r[i,j]-1, level+1);
    FOR k := 1 TO level DO WriteString("      ") END ;
    WriteString(key[r[i,j]]); WriteLn;
    PrintTree(r[i,j], j, level+1)
  END
END PrintTree;

BEGIN (*main program*)
  n := 0; OpenInput("TEXT");
  LOOP ReadCard(b[n]);
    IF NOT Done THEN HALT END ;
    ReadCard(j);
    IF NOT Done THEN EXIT END ;
    n := n+1; a[n] := j;
    ReadString(key[n])
  END ;

  OpenOutput("TREE");
  (*compute w from a and b*)
  FOR i := 0 TO n DO
    w[i,i] := b[i];
    FOR j := i+1 TO n DO
      w[i,j] := w[i,j-1] + a[j] + b[j]
    END
  END ;
  WriteString("Total weight = "); WriteCard(w[0,n], 6); WriteLn;

  WriteString("Pathlength of balanced tree = ");
  WriteCard(BalTree(0, n), 6); WriteLn;
  PrintTree(0, n, 0); WriteLn;

  Read(ch);
```

```
  OptTree;
  WriteString("Pathlength of optimal tree = ");
  WriteCard(p[0,n], 6); WriteLn;
  PrintTree(0, n, 0); WriteLn;

  Read(ch);
  FOR i := 0 TO n DO
   w[i,i] := 0;
   FOR j := i+1 TO n DO
    w[i,j] := w[i,j-1] + a[j]
   END
  END ;
  OptTree;
  WriteString("optimal tree not considering b"); WriteLn;
  PrintTree(0, n, 0); WriteLn;
  CloseInput; CloseOutput
END OptTree.
```

Program 4.6 Find Optimal Search Tree.

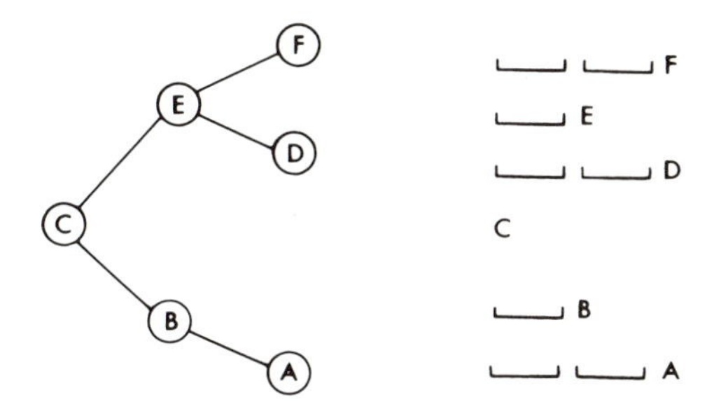

Fig. 4.39 Drawing a tree structure by proper line indentation

b[i-1]	a[i]	k[i]
169	3	AND
25	37	ARRAY
355	125	BEGIN
87	1	BY
264	14	CASE
247	28	CODE
90	9	CONST
118	3	DEFINITION
10	16	DIV
0	55	DO
124	299	ELSE
4	198	ELSIF
10	689	END
281	25	EXIT
35	3	EXPORT
442	19	FROM
0	0	FOR
646	464	IF
5	3	IMPLEMENTATION
13	20	IMPORT
15	2	IN
654	24	LOOP
159	15	MOD
130	16	MODULE
166	79	NIL
16	10	NOT
218	34	OF
31	95	OR
276	11	POINTER
82	171	PROCEDURE
418	1	QUALIFIED
124	6	RECORD
49	9	REPEAT
30	2	RETURN
174	22	SET
505	662	THEN
9	6	TO
385	13	TYPE
37	9	UNTIL
347	203	VAR
84	35	WHILE
0	14	WITH
981		

Table 4.5 Keys and Frequencies of Occurrence

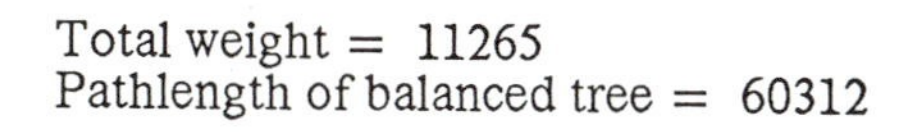

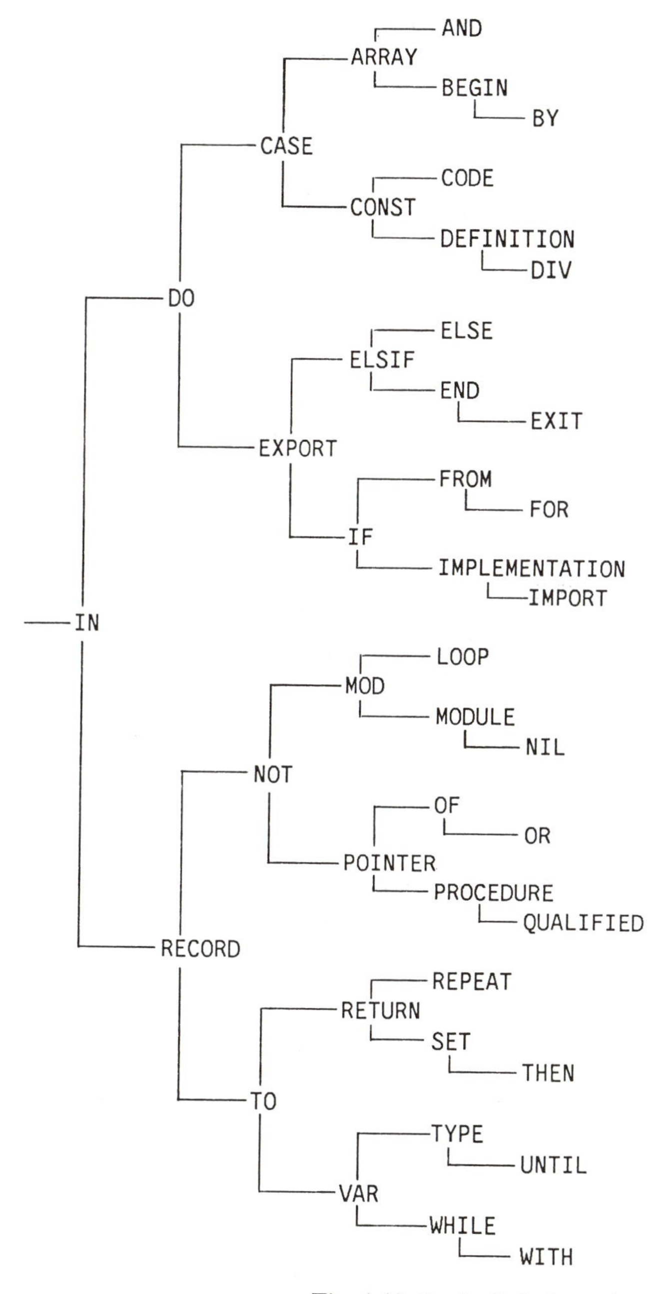

Fig. 4.40 Perfectly balanced tree

Pathlength of optimal tree = 50371

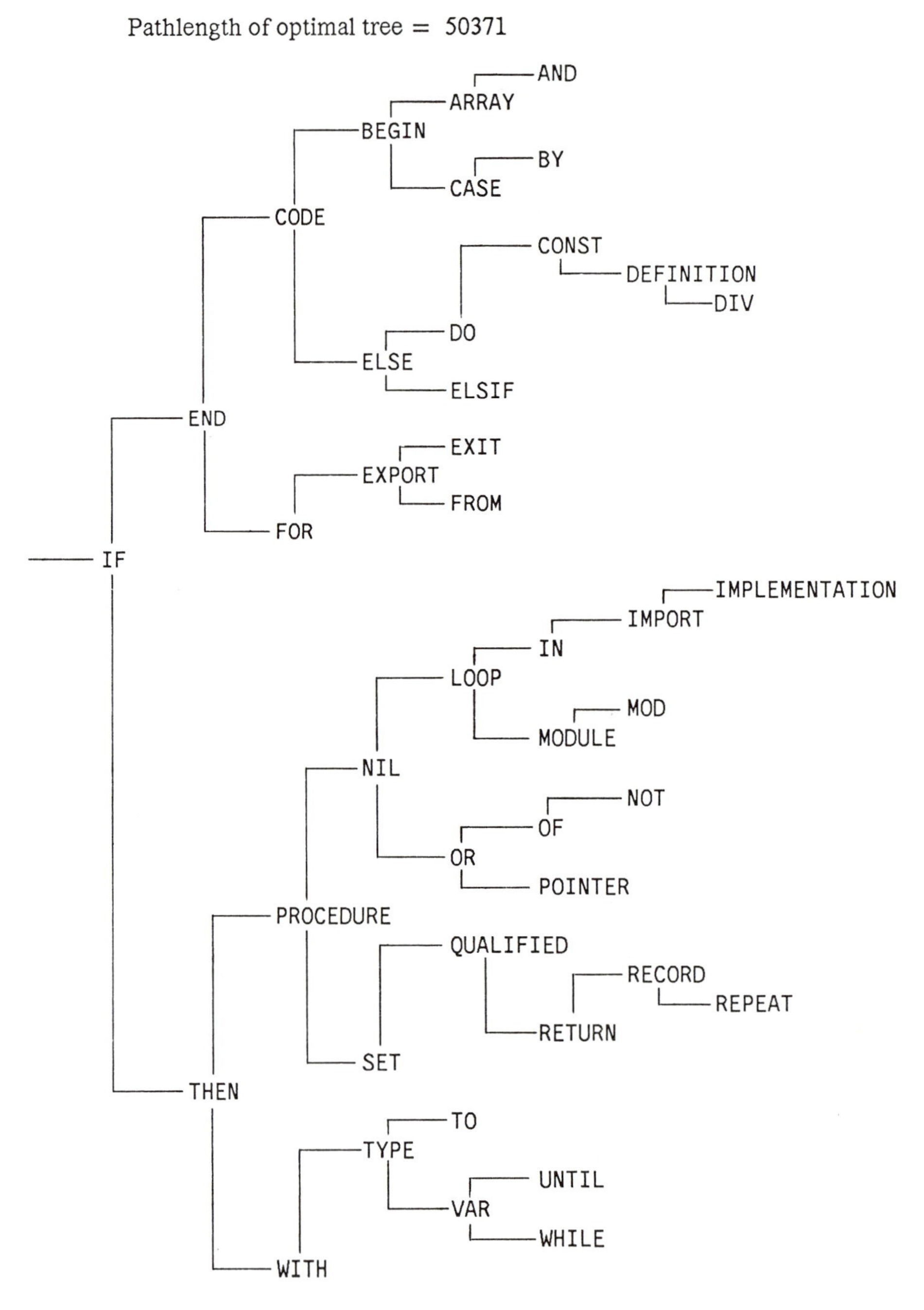

Fig. 4.41 Optimal search tree

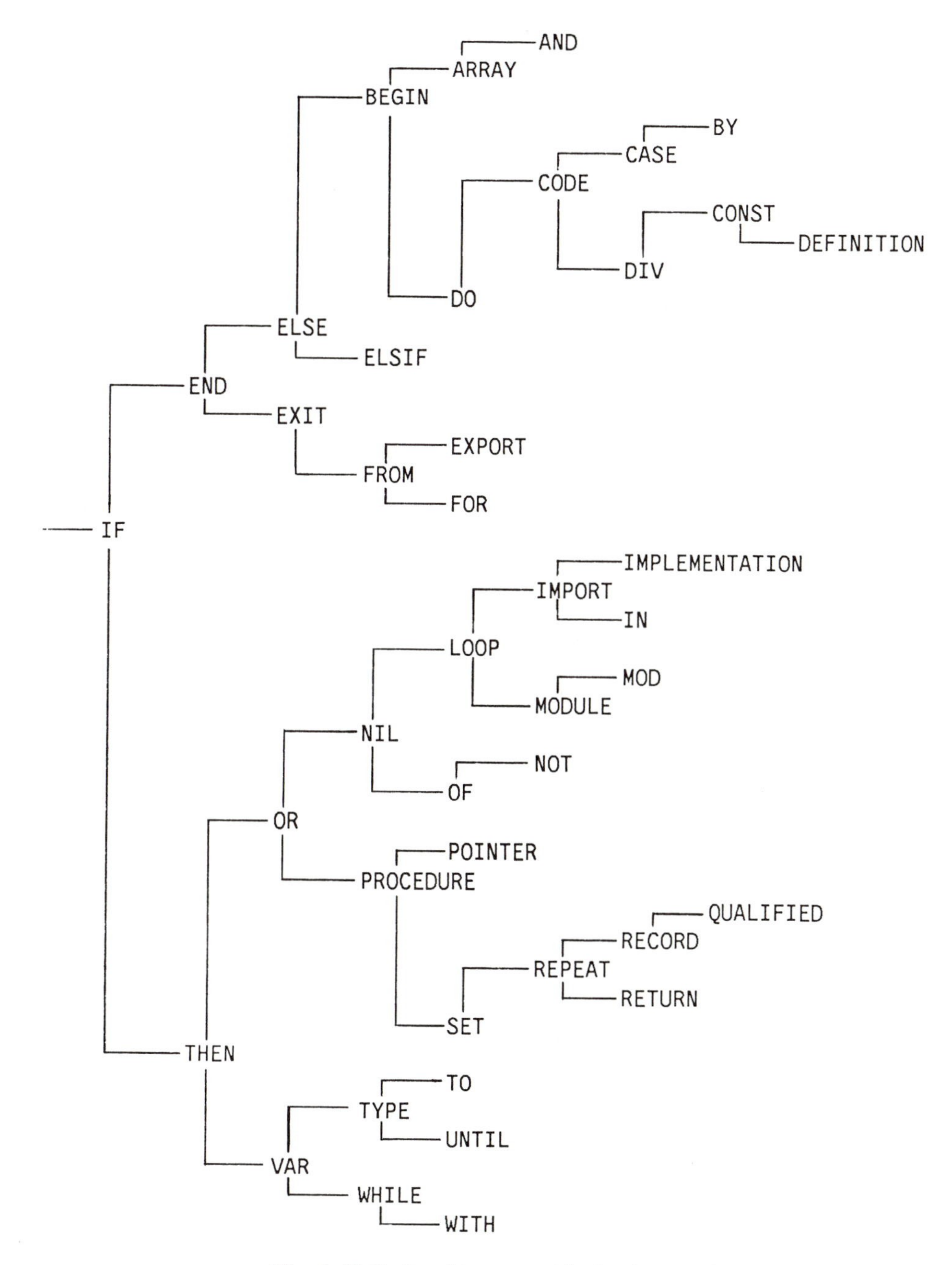

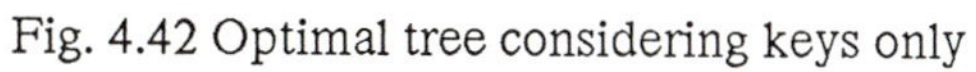
Fig. 4.42 Optimal tree considering keys only

It is evident from algorithm (4.77) that the effort to determine the optimal structure is of the order of n^2; also, the amount of required storage is of the order n^2. This is unacceptable if n is very large. Algorithms with greater efficiency are therefore highly desirable. One of them is the algorithm developed by Hu and Tucker [4-5] which requires only O(n) storage and O(n*log(n)) computations. However, it considers only the case in which the key frequencies are zero, i.e., where only the unsuccessful search trials are registered. Another algorithm, also requiring O(n) storage elements and O(n*log(n)) computations was described by Walker and Gotlieb [4-7]. Instead of trying to find the optimum, this algorithm merely promises to yield a nearly optimal tree. It can therefore be based on heuristic principles. The basic idea is the following.

Consider the nodes (genuine and special nodes) being distributed on a linear scale, weighted by their frequencies (or probabilities) of access. Then find the node which is closest to the center of gravity. This node is called the *centroid*, and its index is

$$((Si: 1 \leq i \leq n : i*a_i) + (Sj: 0 \leq j \leq m : j*b_j)) / W \qquad (4.78)$$

rounded to the nearest integer. If all nodes have equal weight, then the root of the desired optimal tree evidently coincides with the centroid, and -- so the reasoning goes -- it will in most cases be in the close neighborhood of the centroid. A limited search is then used to find the local optimum, whereafter this procedure is applied to the resulting two subtrees. The likelihood of the root lying very close to the centroid grows with the size n of the tree. As soon as the subtrees have reached a manageable size, their optimum can be determined by the above exact algorithm.

4.7. B-TREES

So far, we have restricted our discussion to trees in which every node has at most two descendants, i.e., to binary trees. This is entirely satisfactory if, for instance, we wish to represent family relationships with a preference to the pedigree view, in which every person is associated with his parents. After all, no one has more than two parents. But what about someone who prefers the posterity view? He has to cope with the fact that some people have more than two children, and his trees will contain nodes with many branches. For lack of a better term, we shall call them *multiway trees*.

Of course, there is nothing special about such structures, and we have already encountered all the programming and data definition facilities to cope with such situations. If, for instance, an absolute upper limit on the number of children is given (which is admittedly a somewhat futuristic assumption), then one may represent the children as an array component of the record representing a person. If the number of children varies strongly among different persons, however, this may result in a poor utilization of available storage. In this case it will be much more appropriate to arrange the offspring as a linear list, with a pointer to the youngest (or eldest) offspring assigned to the parent. A possible type definition for this case is (4.80) and a possible data structure is shown in Fig. 4.43.

```
TYPE Ptr =    POINTER TO Person;
TYPE Person = RECORD name: alfa;                          (4.80)
                sibling, offspring: Ptr
              END
```

We now realize that by tilting this picture by 45 degrees it will look like a perfect binary tree. But this view is misleading because functionally the two references have entirely different meanings. One usually dosen't treat a sibling as an offspring and get away unpunished, and hence one should not do so even in constructing data definitions. This example could also be easily extended into an even more complicated data structure by introducing more components in each person's record, thus being able to represent further family relationships. A likely candidate that cannot generally be derived from the sibling and offspring references is that of husband and wife, or even the inverse relationship of father and mother. Such a structure quickly grows into a complex relational data bank, and it may be possible to map serveral trees into it. The algorithms operating on such structures are intimately tied to their data definitions, and it does not make sense to specify any general rules or widely applicable techniques.

However, there is a very practical area of application of multiway trees which is of general interest. This is the construction and maintenance of large-scale search trees in which insertions and deletions are necessary, but in which the primary store of a computer is not large enough or is too costly to be used for long-time storage.

Assume, then, that the nodes of a tree are to be stored on a secondary storage medium such as a disk store. Dynamic data structures introduced in this chapter are particularly suitable for incorporation of secondary storage media. The principal innovation is merely

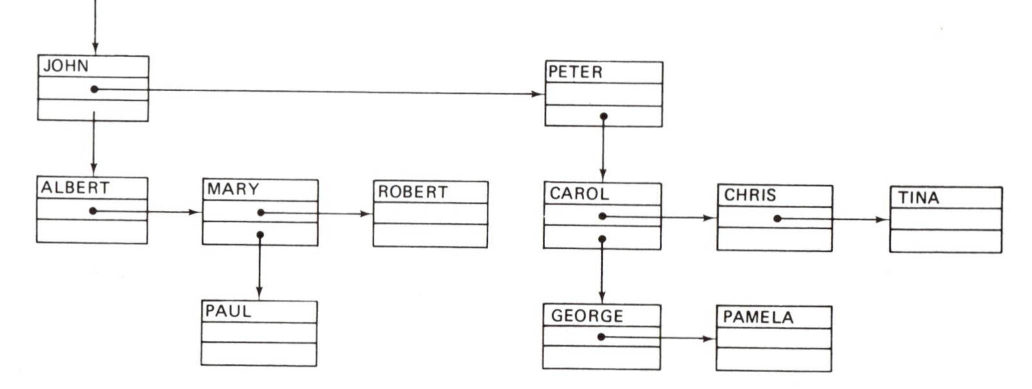

Fig. 4.43 A multiway tree represented as binary tree.

that pointers are represented by disk store addresses instead of main store addresses. Using a binary tree for a data set of, say, a million items, requires on the average approximately log 10^6 (i.e. about 20) search steps. Since each step now involves a disk access (with inherent latency time), a storage organization using fewer accesses will be highly desirable. The multiway tree is a perfect solution to this problem. If an item located on a secondary store is accessed, an entire group of items may also be accessed without much additional cost. This suggests that a tree be subdivided into subtrees, and that the subtrees are represented as units that are accessed all together. We shall call these subtrees *pages*. Figure 4.44 shows a binary tree subdivided into pages, each page consisting of 7 nodes.

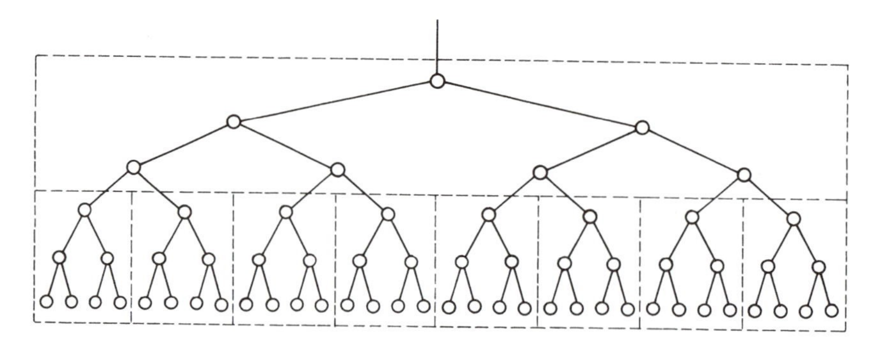

Fig. 4.44 A binary tree subdivided into pages.

The saving in the number of disk accesses -- each page access now involves a disk access -- can be considerable. Assume that we choose to place 100 nodes on a page (this is a reasonable figure); then the million item search tree will on the average require only $\log_{100} 10^6$ (i.e. about 3) page accesses instead of 20. But, of course, if the tree is left to grow at random, then the worst case may still be as large as 10^4. It is plain that a scheme for controlled growth is almost mandatory in the case of multiway trees.

4.7.1. Multiway B-Trees

If one is looking for a controlled growth criterion, the one requiring a perfect balance is quickly eliminated because it involves too much balancing overhead. The rules must clearly be somewhat relaxed. A very sensible criterion was postulated by R. Bayer and E.M. McCreight [4.2] in 1970: every page (except one) contains between n and 2n nodes for a given constant n. Hence, in a tree with N items and a maximum page size of 2n nodes per page, the worst case requires $\log_n N$ page accesses; and page accesses clearly dominate the entire search effort. Moreover, the important factor of store utilization is at least 50% since pages are always at least half full. With all these advantages, the scheme involves comparatively simple algorithms for search, insertion, and deletion. We will subsequently study them in detail.

The underlying data structures are called *B-trees*, and have the following characteristics; n is said to be the *order* of the B-tree.

1. Every page contains at most 2n items (keys.)
2. Every page, except the root page, contains at least n items.
3. Every page is either a leaf page, i.e. has no descendants, or it has m+1 descendants, where m is its number of keys on this page.
4. All leaf pages appear at the same level.

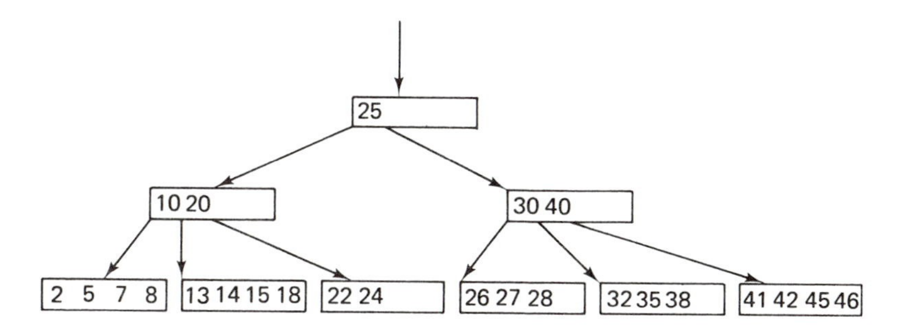

Fig. 4.45 B-tree of order 2.

Figure 4.45 shows a B-tree of order 2 with 3 levels. All pages contain 2, 3, or 4 items; the exception is the root which is allowed to contain a single item only. All leaf pages appear at level 3. The keys appear in increasing order from left to right if the B-tree is squeezed into a single level by inserting the descendants in between the keys of their ancestor page. This arrangement represents a natural extension of binary search trees, and it determines the method of searching an item with given key. Consider a page of the form shown in Fig. 4.46 and a given search argument x. Assuming that the page has been moved into the primary store, we may use conventional search methods among the keys $k_1 \dots k_m$. If m is sufficiently large, one may use binary search; if it is rather small, an ordinary sequential search will do. (Note that the time required for a search in main store is probably negligible compared to the time it takes to move the page from secondary into primary store.) If the search is unsuccessful, we are in one of the following situations:

1. $k_i < x < k_{i+1}$, for $1 \leq i < m$. We continue the search on page $p_i\uparrow$
2. $k_m < x$. The search continues on page $p_m\uparrow$.
3. $x < k_1$. The search continues on page $p_0\uparrow$.

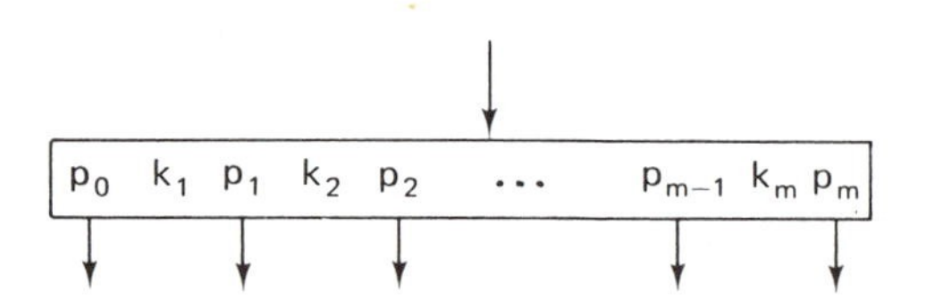

Fig. 4.46 B-tree page with m keys.

If in some case the designated pointer is NIL, i.e., if there is no descendant page, then there is no item with key x in the whole tree, and the search is terminated.

Surprisingly, insertion in a B-tree is comparatively simple too. If an item is to be inserted in a page with $m < 2n$ items, the insertion process remains constrained to that page. It is only insertion into an already full page that has consequences upon the tree structure and may cause the allocation of new pages. To understand what happens in this case, refer to Fig. 4.47, which illustrates the insertion of key 22 in a B-tree of order 2. It proceeds in the following steps:

1. Key 22 is found to be missing; insertion in page C is impossible because C is already full.
2. Page C is *split* into two pages (i.e., a new page D is allocated).
3. The $2n+1$ keys are equally distributed onto C and D, and the middle key is moved up one level into the ancestor page A.

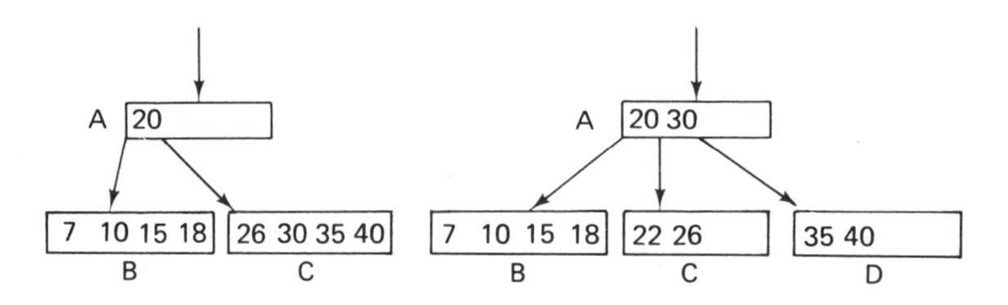

Fig. 4.47 Insertion of key 22 in B-tree.

This very elegant scheme preserves all the characteristic properties of B-trees. In particular, the split pages contain exactly n items. Of course, the insertion of an item in the ancestor page may again cause that page to *overflow,* thereby causing the splitting to propagate. In the extreme case it may propagate up to the root. This is, in fact, the only way that the B-tree may increase its height. The B-tree has thus a strange manner of growing: it grows from its leaves upward to the root.

We shall now develop a detailed program from these sketchy descriptions. It is already apparent that a recursive formulation will be most convenient because of the property of the splitting process to propagate back along the search path. The general structure of the program will therefore be similar to balanced tree insertion, although the details are different. First of all, a definition of the page structure has to be formulated. We choose to represent the items in the form of an array.

```
TYPE PPtr =  POINTER TO Page;
TYPE index = [0 .. 2*n];

TYPE item =  RECORD key: INTEGER;                         (4.81)
               p: PPtr;
               count: CARDINAL
             END ;

TYPE page =  RECORD m: index;                             (4.82)
               p0: PPtr;
               e: ARRAY [1 .. 2*n] OF item
             END
```

Again, the item component *count* stands for all kinds of other information that may be associated with each item, but it plays no role in the actual search process. Note that each page offers space for 2n items. The field m indicates the actual number of items on the page. As $m \geq n$ (except for the root page), a storage utilization of a least 50% is guaranteed.

The algorithm of B-tree search and insertion is part of Program 4.7, formulated as a procedure called *search.* Its main structure is straightforward and similar to that for the balanced binary tree search, with the exception that the branching decision is not a binary choice. Instead, the *within-page search* is represented as a binary search on the array e of elements.

The insertion algorithm is formulated as a separate procedure merely for clarity. It is activated after search has indicated that an item is to be passed up on the tree (in the direction toward the root). This fact is indicated by the Boolean result parameter h; it assumes a similar role as in the algorithm for balanced tree insertion, where h indicates that the subtree had grown. If h is true, the second result parameter, u, represents the item being passed up. Note that insertions start in hypothetical pages, namely, the "special nodes" of Fig. 4.19; the new item is immediately handed up via the parameter u to the leaf page for actual insertion. The scheme is sketched in (4.83).

```
PROCEDURE search(x: INTEGER; a: PPtr; VAR h: BOOLEAN; VAR u: item);
```

```
BEGIN
  IF a = NIL THEN (*x not in tree, insert*)                        (4.83)
    Assign x to item u, set h to TRUE, indicating that an item
    u is passed up in the tree
  ELSE
    WITH a↑ DO
      binary search for x in array e;
      IF found THEN process data
      ELSE search(x, descendant, h, u);
        IF h THEN (*an item was passed up*)
          IF no. of items on page a↑ < 2n THEN
            insert u on page a↑ and set h to FALSE
          ELSE split page and pass middle item up
          END
        END
      END
    END
  END
END search
```

If the paramerter h is true after the call of *search* in the main program, a split of the root page is requested. Since the root page plays an exceptional role, this process has to be programmed separately. It consists merely of the allocation of a new root page and the insertion of the single item given by the paramerter u. As a consequence, the new root page contains a single item only. The details can be gathered from Program 4.7, and Fig. 4.48 shows the result of using Program 4.7 to construct a B-tree with the following insertion sequence of keys:

20; 40 10 30 15; 35 7 26 18 22; 5; 42 13 46 27 8 32; 38 24 45 25;

The semicolons designate the positions of the snapshots taken upon each page allocation. Insertion of the last key causes two splits and the allocation of three new pages.

The with clause in this program has a special significance. In the first place it indicates that identifiers of page components automatically refer to page a↑ within the statement prefixed by the clause. If, in fact, the pages are allocated on secondary store -- as would certainly be necessary in a large data base system -- then the with clause may in addition be interpreted as implying the transfer of the designated page into primary store. Since each activation of *search* therefore implies one page transfer to main store, $k = \log_n N$ recursive calls are necessary at most, if the tree contains N items. Hence, we must be capable of accommodating k pages in main store. This is one limiting factor on the page size 2n. In fact, we need to accommodate even more than k pages, because insertion may cause page splitting to occur. A corollary is that the root page is best allocated permanently in the primary store, because each query proceeds necessarily through the root page.

Another positive quality of the B-tree organization is its suitability and economy in the case of purely *sequential updating* of the entire data base. Every page is fetched into primary store exactly once.

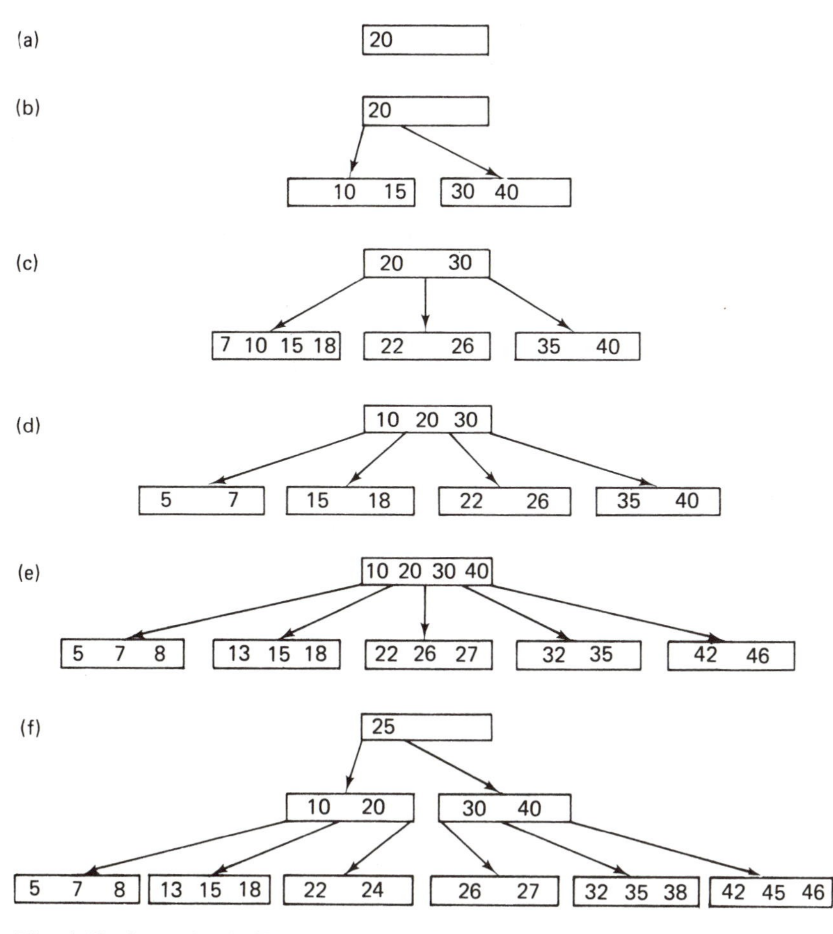

Fig. 4.48 Growth of a B-tree of order 2.

Deletion of items from a B-tree is fairly straight-forward in principle, but it is complicated in the details. We may distinguish two different circumstances:

1. The item to be deleted is on a leaf page; here its removal algorithm is plain and simple.

2. The item is not on a leaf page; it must be replaced by one of the two lexicographically adjacent items, which happen to be on leaf pages and can easily be deleted.

In case 2 finding the adjacent key is analogous to finding the one used in binary tree deletion. We descend along the rightmost pointers down to the leaf page P, replace the item to be deleted by the rightmost item on P, and then reduce the size of P by 1. In any case, reduction of size must be followed by a check of the number of items m on the reduced page, because, if m < n, the primary characteristic of B-trees would be violated. Some additional action has to be taken; this *underflow* condition is indicated by the Boolean variable parameter h.

The only recourse is to borrow or annect an item from one of the neighboring pages, say from Q. Since this involves fetching page Q into main store -- a relatively costly operation -- one is tempted to make the best of this undesirable situation and to annect more than a single item at once. The usual strategy is to distribute the items on pages P and Q evenly on both pages. This is called *page balancing.*

Of course, it may happen that there is no item left to be annected since Q has already reached its minimal size n. In this case the total number of items on pages P and Q is 2n-1; we may *merge* the two pages into one, adding the middle item from the ancestor page of P and Q, and then entirely dispose of page Q. This is exactly the inverse process of page splitting. The process may be visualized by considering the deletion of key 22 in Fig. 4.47. Once again, the removal of the middle key in the ancestor page may cause its size to drop below the permissible limit n, thereby requiring that further special action (either balancing or merging) be undertaken at the next level. In the extreme case page merging may propagate all the way up to the root. If the root is reduced to size 0, it is itself deleted, thereby causing a reduction in the height of the B-tree. This is, in fact, the only way that a B-tree may shrink in height. Figure 4.49 shows the gradual decay of the B-tree of Fig. 4.48 upon the sequential deletion of the keys

25 45 24; 38 32; 8 27 46 13 42; 5 22 18 26; 7 35 15;

The semicolons again mark the places where the snapshots are taken, namely where pages are being eliminated. The deletion algorithm is included as a procedure in Program 4.7. The similarity of its structure to that of balanced tree deletion is particularly noteworthy.

```
MODULE BTree;
  FROM InOut IMPORT OpenInput, OpenOutput, CloseInput, CloseOutput,
     ReadInt, Done, Write, WriteInt, WriteString, WriteLn;
  FROM Storage IMPORT ALLOCATE;

  CONST n = 2;

  TYPE PPtr = POINTER TO Page;

   Item = RECORD key: INTEGER;
        p: PPtr;
        count: CARDINAL
      END ;

   Page = RECORD m: [0 .. 2*n];  (*no. of items on page*)
        p0: PPtr;
        e: ARRAY [1 .. 2*n] OF Item
      END ;

  VAR root, q: PPtr;
    x: INTEGER;
    h: BOOLEAN;
    u: Item;
```

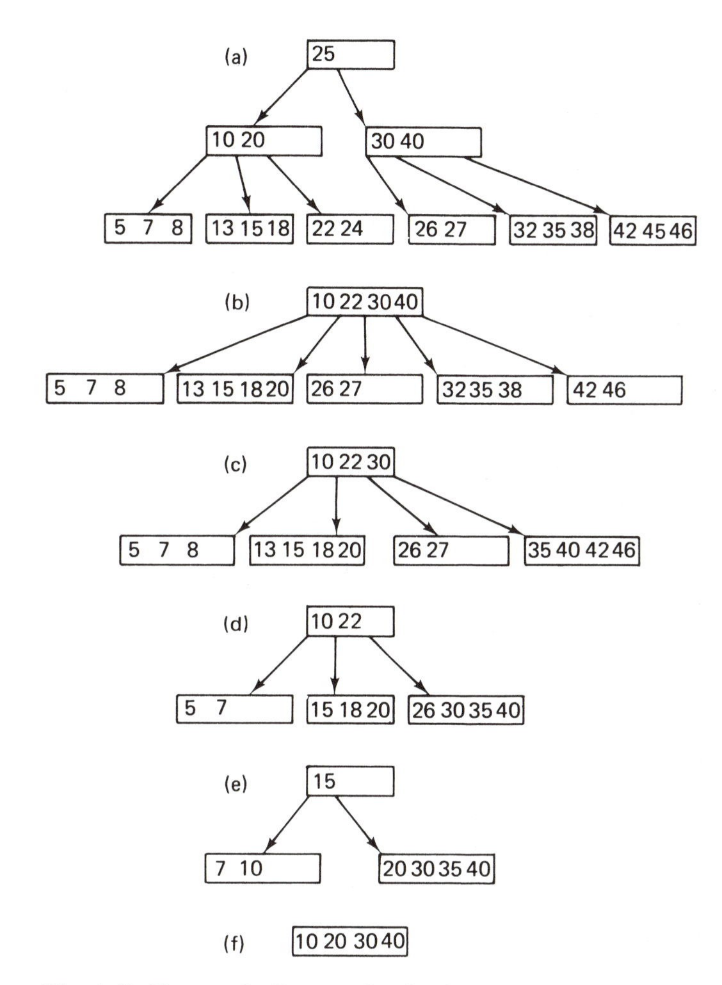

Fig. 4.49 Decay of a B-tree of order 2.

```
PROCEDURE search(x: INTEGER; a: PPtr; VAR h: BOOLEAN; VAR v: Item);
  (*search key x in B-tree with root a; if found, increment counter.
    Otherwise insert new item with key x. If an item is passed up,
    assign it to v. h = "tree has become higher"*)
  VAR i, L, R: CARDINAL; b: PPtr; u: Item;
BEGIN (*~h*)
  IF a = NIL THEN h := TRUE; (*not in tree*)
    WITH v DO
      key := x; count := 1; p := NIL
    END
  ELSE
    WITH a↑ DO
      L := 1; R := m+1; (*binary search*)
```

```
        WHILE L < R DO
         i := (L+R) DIV 2;
         IF e[i].key <= x THEN L := i+1 ELSE R := i END
        END ;
        R := R-1;
        IF (R > 0) & (e[R].key = x) THEN INC(e[R].count)
        ELSE (*item not on this page*)
         IF R = 0 THEN search(x, p0, h, u)
          ELSE search(x, e[R].p, h, u)
         END ;
         IF h THEN (*insert u to the right of e[R]*)
          IF m < 2*n THEN
           h := FALSE; m := m+1;
           FOR i := m TO R+2 BY -1 DO e[i] := e[i-1] END ;
           e[R+1] := u
          ELSE ALLOCATE(b, SIZE(Page)); (*overflow*)
           (*split a into a,b and assign the middle item to v*)
           IF R <= n THEN
            IF R = n THEN v := u
            ELSE v := e[n];
             FOR i := n TO R+2 BY -1 DO e[i] := e[i-1] END ;
             e[R+1] := u
            END ;
            FOR i := 1 TO n DO b↑.e[i] := a↑.e[i+n] END
           ELSE (*insert in right page*)
            R := R-n; v := e[n+1];
            FOR i := 1 TO R-1 DO b↑.e[i] := a↑.e[i+n+1] END ;
            b↑.e[R] := u;
            FOR i := R+1 TO n DO b↑.e[i] := a↑.e[i+n] END
           END ;
           m := n; b↑.m := n; b↑.p0 := v.p; v.p := b
          END
         END
        END
      END
    END
  END search;

  PROCEDURE underflow(c, a: PPtr; s: CARDINAL; VAR h: BOOLEAN);
   (*a = underflowing page, c = ancestor page,
    s = index of deleted item in c, h := *)
   VAR b: PPtr; VAR i, k, mb, mc: CARDINAL;
  BEGIN mc := c↑.m; (*h, a↑.m = n-1*)
   IF s < mc THEN
    (*b := page to the right of a*) s := s+1;
    b := c↑.e[s].p; mb := b↑.m; k := (mb-n+1) DIV 2;
```

```
  (*k = no. of items available on page b*)
  a↑.e[n] := c↑.e[s]; a↑.e[n].p := b↑.p0;
  IF k > 0 THEN
   (*move k items from b to a*)
   FOR i := 1 TO k-1 DO a↑.e[i+n] := b↑.e[i] END ;
   c↑.e[s] := b↑.e[k]; c↑.e[s].p := b;
   b↑.p0 := b↑.e[k].p; mb := mb - k;
   FOR i := 1 TO mb DO b↑.e[i] := b↑.e[i+k] END ;
   b↑.m := mb; a↑.m := n-1+k; h := FALSE
  ELSE (*merge pages a and b*)
   FOR i := 1 TO n DO a↑.e[i+n] := b↑.e[i] END ;
   FOR i := s TO mc-1 DO c↑.e[i] := c↑.e[i+1] END ;
   a↑.m := 2*n; c↑.m := mc-1; h := mc <= n;
   (*Deallocate(b)*)
  END
 ELSE (*b := page to the left of a*)
  IF s = 1 THEN b := c↑.p0 ELSE b := c↑.e[s-1].p END ;
  mb := b↑.m + 1; k := (mb-n) DIV 2;
  IF k > 0 THEN
   (*move k items from page b to a*)
   FOR i := n-1 TO 1 BY -1 DO a↑.e[i+k] := a↑.e[i] END ;
   a↑.e[k] := c↑.e[s]; a↑.e[k].p := a↑.p0; mb := mb-k;
   FOR i := k-1 TO 1 BY -1 DO a↑.e[i] := b↑.e[i+mb] END ;
   a↑.p0 := b↑.e[mb].p; c↑.e[s] := b↑.e[mb]; c↑.e[s].p := a;
   b↑.m := mb-1; a↑.m := n-1+k; h := FALSE
  ELSE (*merge pages a and b*)
   b↑.e[mb] := c↑.e[s]; b↑.e[mb].p := a↑.p0;
   FOR i := 1 TO n-1 DO b↑.e[i+mb] := a↑.e[i] END ;
   b↑.m := 2*n; c↑.m := mc-1; h := mc <= n;
   (*Deallocate(a)*)
  END
 END
END underflow;

PROCEDURE delete(x: INTEGER; a: PPtr; VAR h: BOOLEAN);
 (*search and delete key x in B-tree a; if a page underflow arises,
  balance with adjacent page or merge; h := "page a is undersize"*)
 VAR i, L, R: CARDINAL; q: PPtr;

 PROCEDURE del(P: PPtr; VAR h: BOOLEAN);
  VAR q: PPtr; (*global a, R*)
 BEGIN
  WITH P↑ DO
   q := e[m].p;
   IF q # NIL THEN del(q,h);
    IF h THEN underflow(P, q, m, h) END
```

```
      ELSE
        P↑.e[m].p := a↑.e[R].p; a↑.e[R] := P↑.e[m];
        m := m - 1; h := m < n
      END
    END
  END del;

BEGIN
  IF a = NIL THEN (*x not in tree*) h := FALSE
  ELSE
    WITH a↑ DO
      L := 1; R := m+1;  (*binary search*)
      WHILE L < R DO
        i := (L+R) DIV 2;
        IF e[i].key < x THEN L := i+1 ELSE R := i END
      END ;
      IF R = 1 THEN q := p0 ELSE q := e[R-1].p END ;
      IF (R <= m) & (e[R].key = x) THEN
        (*found, now delete*)
        IF q = NIL THEN  (*a is a terminal page*)
          m := m-1; h := m < n;
          FOR i := R TO m DO e[i] := e[i+1] END
        ELSE del(q,h);
          IF h THEN underflow(a, q, R-1, h) END
        END
      ELSE delete(x, q, h);
        IF h THEN underflow(a, q, R-1, h) END
      END
    END
  END
END delete;

PROCEDURE PrintTree(p: PPtr; level: CARDINAL);
  VAR i: CARDINAL;
BEGIN
  IF p # NIL THEN
    FOR i := 1 TO level DO WriteString("      ") END ;
    FOR i := 1 TO p↑.m DO WriteInt(p↑.e[i].key, 4) END ;
    WriteLn;
    PrintTree(p↑.p0, level+1);
    FOR i := 1 TO p↑.m DO PrintTree(p↑.e[i].p, level+1) END
  END
END PrintTree;

BEGIN (*main program*)
  OpenInput("TEXT"); OpenOutput("TREE");
  root := NIL; Write(">"); ReadInt(x);
```

```
  WHILE Done DO
   WriteInt(x, 5); WriteLn;
   IF x >= 0 THEN
    search(x, root, h, u);
    IF h THEN (*insert new base page*)
     q := root; ALLOCATE(root, SIZE(Page));
     WITH root↑ DO
      m := 1; p0 := q; e[1] := u
     END
    END
   ELSE
    delete(-x, root, h);
    IF h THEN (*base page size reduced*)
     IF root↑.m = 0 THEN
      q := root; root := q↑.p0; (*Deallocate(q)*)
     END
    END
   END ;
   PrintTree(root, 0); WriteLn;
   Write(">"); ReadInt(x)
  END ;
  CloseInput; CloseOutput
 END BTree.
```

Program 4.7 B-Tree Search, Insertion, and Deletion.

Extensive analysis of B-tree performance has been undertaken and is reported in the referenced article (Bayer and McCreight). In particular, it includes a treatment of the question of optimal page size, which strongly depends on the characteristics of the storage and computing system available.

Variations of the B-tree scheme are discussed in Knuth, Vol. 3, pp. 476-479. The one notable observation is that page splitting should be delayed in the same way that page merging is delayed, by first attempting to balance neighboring pages. Apart from this, the suggested improvements seem to yield marginal gains. A comprehensive survey of B-trees may be found in [4-8].

4.7.2. Binary B-Trees

The species of B-trees that seems to be least interesting is the first order B-tree ($n = 1$). But sometimes it is worthwhile to pay attention to the exceptional case. It is plain, however, that first-order B-trees are not useful in representing large, ordered, indexed data sets invoving secondary stores; approximately 50% of all pages will contain a single item only. Therefore, we shall forget secondary stores and again consider the problem of search trees involving a one-level store only.

A *binary B-tree*(BB-tree) consists of nodes (pages) with either one or two items. Hence, a page contains either two or three pointers to descendants; this suggested the term *2-3 tree.*

According to the definition of B-trees, all leaf pages appear at the same level, and all non-leaf pages of BB-trees have either two or three descendants (including the root). Since we now are dealing with primary store only, an optimal economy of storage space is mandatory, and the representation of the items inside a node in the form of an array appears unsuitable. An alternative is the dynamic, linked allocation; that is, inside each node there exists a linked list of items of length 1 or 2. Since each node has at most three descendants and thus needs to harbor only up to three pointers, one is tempted to combine the pointers for descendants and pointers in the item list as shown in Fig. 4.50. The B-tree node thereby loses its actual identity, and the items assume the role of nodes in a regular binary tree. It remains necessary, however, to distinguish between pointers to descendants (vertical) and pointers to siblings on the same page (horizontal). Since only the pointers to the right may be horizontal, a single bit is sufficient to record this distiction. We therefore introduce the Boolean field h with the meaning *horizontal*. The definition of a tree node based on this representation is given in (4.84). It was suggested and investigated by R. Bayer [4-3] in 1971 and represents a search tree organization guaranteeing a maximum path length $p = 2*\lceil \log N \rceil$.

```
TYPE Ptr =   POINTER TO Node;
TYPE Node = RECORD key: INTEGER;                                    (4.84)
               ..........
               left, right: Ptr;
               h: BOOLEAN  (*right branch horizontal*)
             END
```

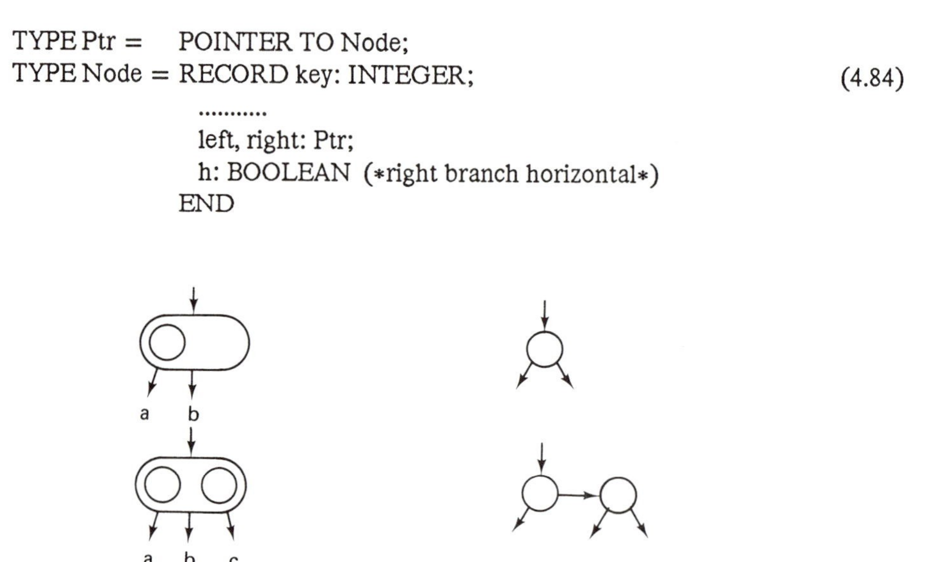

Fig. 4.50 Representation of BB-tree nodes.

Considering the problem of key insertion, one must distinguish four possible situations that arise from growth of the left or right subtrees. The four cases are illustrated in Fig. 4.51. Remember that B-trees have the characteristic of growing from the bottom toward the root and that the property of all leafs being at the same level must be maintained. The simplest case (1) is when the right subtree of a node A grows and when A is the only key on its (hypothetical) page. Then, the descendant B merely becomes the sibling of A, i.e., the vertical pointer becomes a horizontal pointer. This simple raising of the right arm is not possible if A already has a sibling. Then we would obtain a page with 3 nodes, and we have to split it (case 2). Its middle node B is passed up to the next higher level.

Now assume that the left subtree of a node B has grown in height. If B is again alone on a page (case 3), i.e., its right pointer refers to a descentant, then the left subtree (A) is allowed to become B's sibling. (A simple rotation of pointers is necessary since the left pointer cannot be horizontal.) If, however, B already has a sibling, the raising of A yields a page with three members, requiring a split. This split is realized in a very straightforward manner: C becomes a descendant of B, which is raised to the next higher level (case 4).

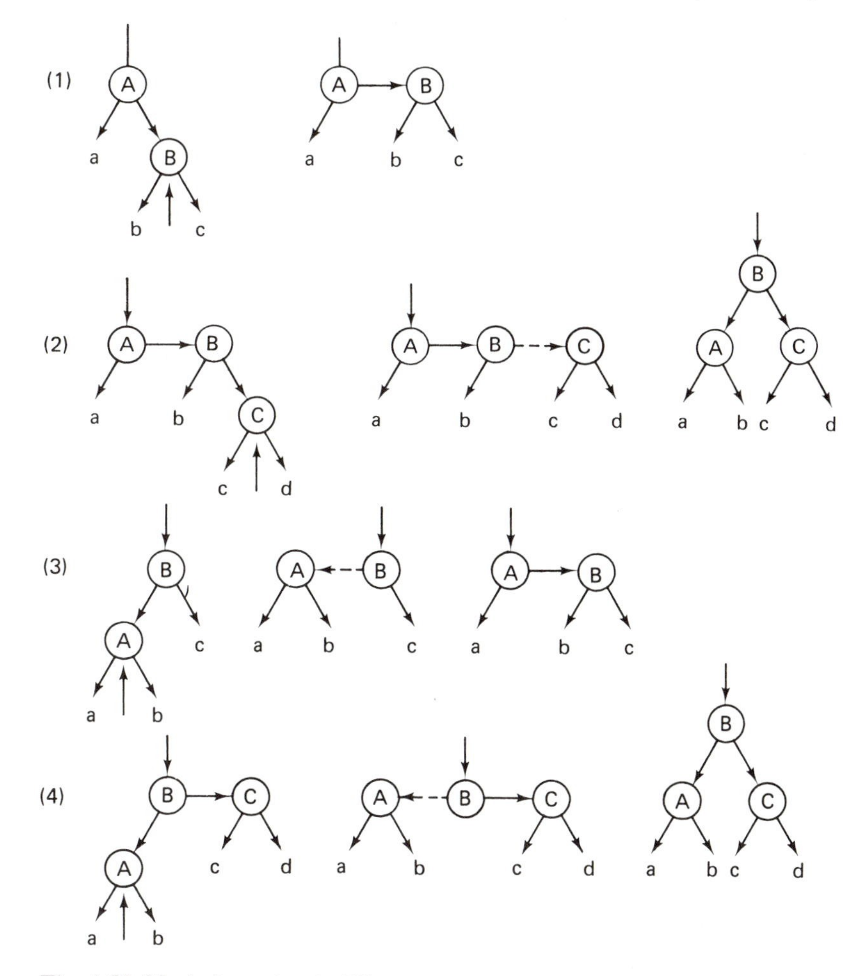

Fig. 4.51 Node insertion in BB-tree.

It should be noted that upon searching a key, it makes no effective difference whether we proceed along a horizontal or a vertical pointer. It therefore appears artificial to worry about a left pointer in case 3 becoming horizontal, although its page still contains not more than two members. Indeed, the insertion algorithm reveals a strange asymmetry in handling the growth of left and right subtrees, and it lets the BB-tree organization appear rather artificial. There is no proof of strangeness of this organization; yet a healthy intuition tells us that something is fishy, and that we should remove this asymmetry. It leads to the notion of the *symmetric binary B-tree*(SBB-tree) which was also investigated by Bayer [4-4] in 1972. On the average it leads to slightly more efficient search trees, but the algorithms for insertion and deletion are also slightly more complex. Furthermore, each node now requires two bits (Boolean variable lh and rh) to indicate the nature of its two pointers.

Since we will restrict our detail considerations to the problem of insertion, we have once again to distinguish among four cases of grown subtrees. They are illustrated in Fig. 4.52, which makes the gained symmetry evident. Note that whenever a subtree of node A without siblings grows, the root of the subtree becomes the sibling of A. This case need not be considered any further.

The four cases considered in Fig. 4.52 all reflect the occurrence of a page overflow and the subsequent page split. They are labelled according to the directions of the horizontal pointers linking the three siblings in the middle figures. The initial situation is shown in the left column; the middle column illustrates the fact that the lower node has been raised as its subtree has grown; the figures in the right column show the result of node rearrangement.

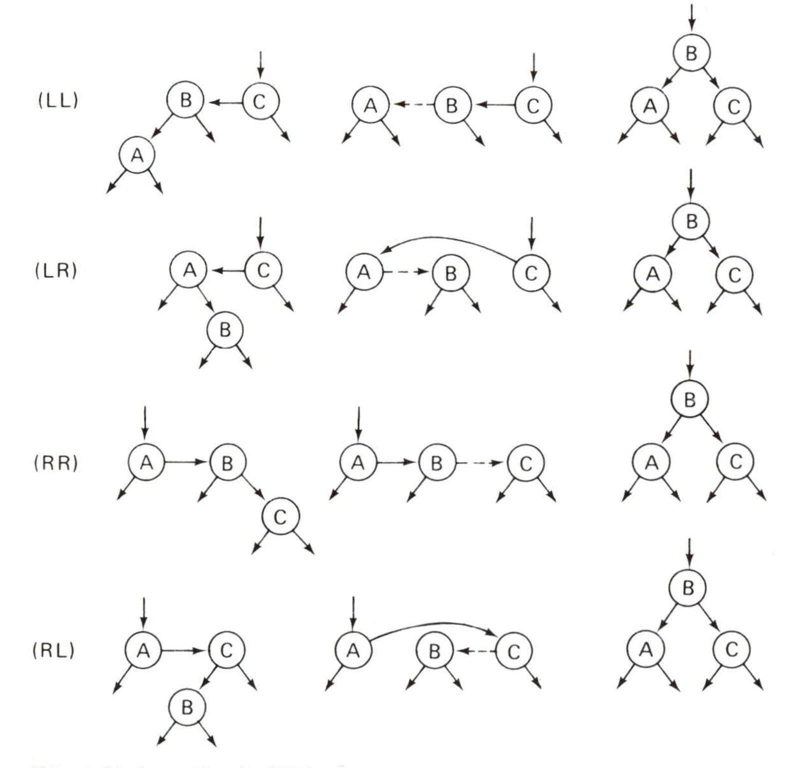

Fig. 4.52 Insertion in SBB-trees.

It is advisable to stick no longer to the notion of pages out of which this organization had developed, for we are only interested in bounding the maximum path length to 2*log N. For this we need only ensure that two horizontal pointers may never occur in succession on any search path. However, there is no reason to forbid any nodes with horizontal pointers to the left and right. We therefore define the SBB-tree as a tree that has the following properties:

1. Every node contains one key and at most two (pointers to) subtrees.
2. Every pointer is either horizontal or vertical. There are no two consecutive horizontal pointers on any search path.
3. All terminal nodes (nodes without descendants) appear at the same (terminal) level.

From this definition it follows that the longest search path is no longer than twice the height of the tree. Since no SBB-tree with N nodes can have a height larger than log N, it follows immediately that $2*\lceil \log N \rceil$ is an upper bound on the search path length. In order to visualize how these trees grow, we refer to Fig. 4.53. The lines represent snapshots taken during the insertion of the following sequences of keys, where every semicolon marks a snapshot.

```
(1)  1  2;  3;  4  5  6;  7;
(2)  5  4;  3;  1  2  7  6;          (4.85)
(3)  6  2;  4;  1  7  3  5;
(4)  4  2  6;  1  7;  3  5;
```

These pictures make the third property of B-trees particularly obvious: all terminal nodes appear on the same level. One is therefore inclined to compare these structures with garden hedges that have been recently trimmed with hedge scissors.

The algorithm for the construction of SBB-trees is formulated in (4.87). It is based on a definition of the type *Node* (4.86) with the two components *lh* and *rh* indicating whether or not the left and right pointers are horizontal.

```
TYPE Node = RECORD key: integer;
                count: CARDINAL;
                left, right: Ptr;                    (4.86)
                lh, rh: BOOLEAN
              END
```

The recursive procedure *search* again follows the pattern of the basic binary tree insertion algorithm (see 4.87). A third parameter h is added; it indicates whether or not the subtree with root p has changed, and it corresponds directly to the parameter h of the B-tree search program. We must note, however, the consequence of representing pages as linked lists: a page is traversed by either one or two calls of the search procedure. We must distinguish between the case of a subtree (indicated by a vertical pointer) that has grown and a sibling node (indicated by a horizontal pointer) that has obtained another sibling and hence requires a page split. The problem is easily solved by introducing a three-valued h with the following meanings:

1. h = 0: the subtree p requires no changes of the tree structure.
2. h = 1: node p has obtained a sibling.
3. h = 2: the subtree p has increased in height.

```
PROCEDURE search(x: INTEGER; VAR p: Ptr; VAR h: CARDINAL);
  VAR p1, p2: Ptr; (*h = 0*)
BEGIN                                                        (4.87)
```

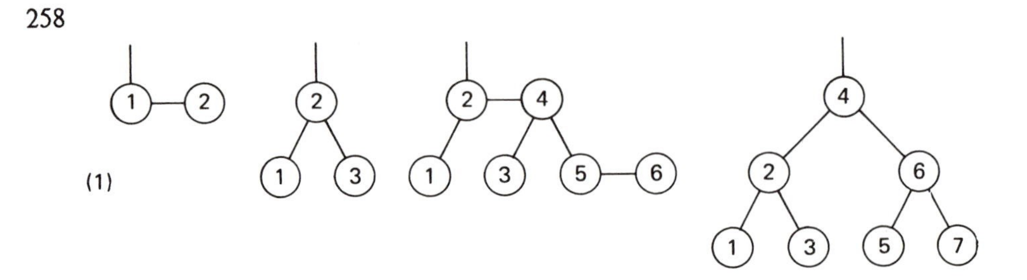

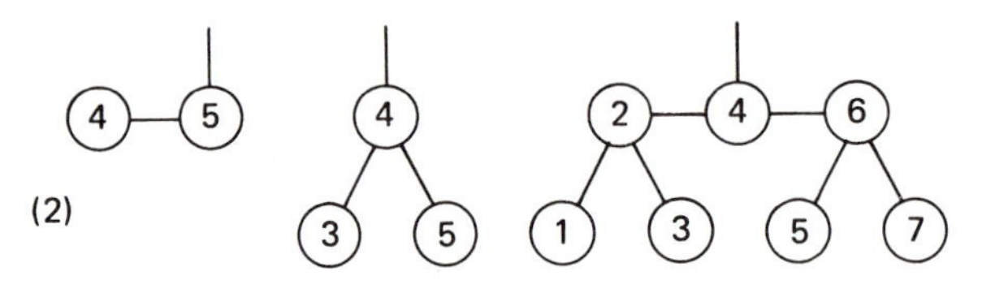

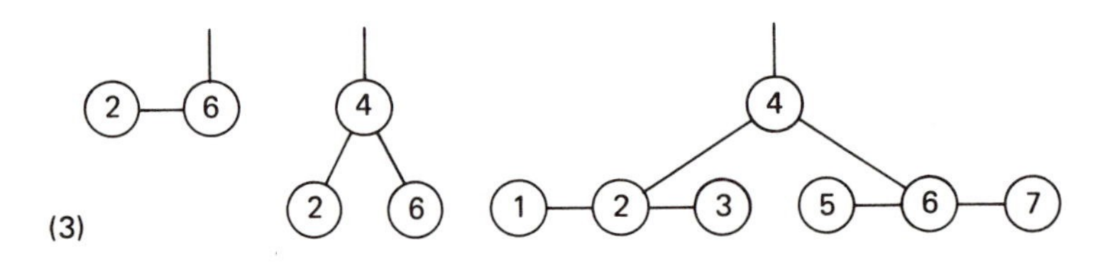

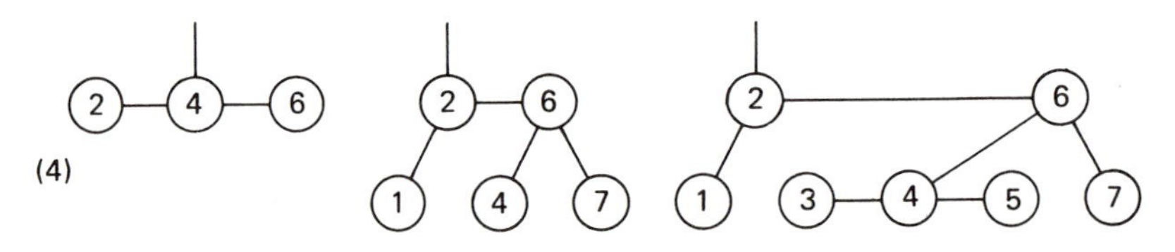

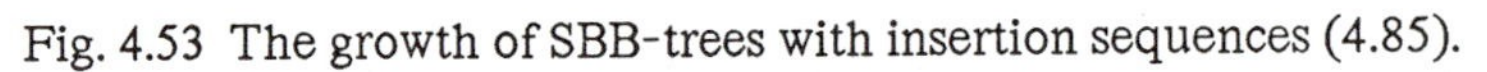

Fig. 4.53 The growth of SBB-trees with insertion sequences (4.85).

```
IF p = NIL THEN (*insert*)
  ALLOCATE(p, SIZE(Node)); h := 2;
  WITH p↑ DO
    key := x; count := 1;
    left := NIL; right := NIL; lh := FALSE; rh := FALSE
  END
ELSIF p↑.key > x THEN
  search(x, p↑.left, h);
  IF h > 0 THEN (*left branch has grown*)
    IF p↑.lh THEN
      p1 := p↑.left; h := 2; p↑.lh := FALSE;
      IF p1↑.lh THEN (*LL*)
        p↑.left := p1↑.right; p1↑.right := p; p := p1;
        p↑.lh := FALSE
```

```
      ELSIF p1↑.rh THEN (*LR*)
        p2 := p1↑.right; p1↑.right := p2↑.left; p2↑.left := p1;
        p↑.left := p2↑.right; p2↑.right := p; p := p2;
        p1↑.rh := FALSE
      END
    ELSE h := h-1;
      IF h > 0 THEN p↑.lh := TRUE END
    END
  END
ELSIF p↑.key < x THEN
  search(x, p↑.right, h);
  IF h > 0 THEN (*right branch has grown*)
    IF p↑.rh THEN
      p1 := p↑.right; h := 2; p↑.rh := FALSE;
      IF p1↑.rh THEN (*RR*)
        p↑.right := p1↑.left; p1↑.left := p; p := p1;
        p1↑.rh := FALSE
      ELSIF p1↑.lh THEN (*RL*)
        p2 := p1↑.left; p1↑.left := p2↑.right; p2↑.right := p1;
        p↑.right := p2↑.left; p2↑.left := p; p := p2;
        p1↑.lh := FALSE
      END
    ELSE h := h-1;
      IF h > 0 THEN p↑.rh := TRUE END
    END
  END
ELSE (*found*) p↑.count := p↑.count + 1
END
END search
```

Note that the actions to be taken for node rearrangement very strongly resemble those developed in the balanced tree search algorithm (4.63). From (4.87) it is evident that all four cases can be implemented by simple pointer rotations: single rotations in the LL and RR cases, double rotations in the LR and RL cases. In fact, procedure (4.87) appears slightly simpler than (4.63). Clearly, the SBB-tree scheme emerges as an alternative to the AVL-balance criterion. A performance comparison is therefore both possible and desirable.

We refrain from involved mathematical analysis and concentrate on some basic differences. It can be proven that the AVL-balanced trees are a subset of the SBB-trees. Hence, the class of the latter is larger. It follows that their path length is on the average larger than in the AVL case. Note in this connection the worst-case tree (4) in Fig. 4.53. On the other hand, node rearrangement is called for less frequently. The balanced tree is therefore preferred in those applications in which key retrievals are much more frequent than insertions (or deletions); if this quotient is moderate, the SBB-tree scheme may be preferred. It is very difficult to say where the borderline lies. It strongly depends not only on the quotient between the frequencies of retrieval and structural change, but also on the characteristics of an implementation. This is particularly the case if the node records have a densely packed representation, and if therefore access to fields involves part-word selection.

4.8 PRIORITY SEARCH TREES

Trees, and in particular binary trees, constitute very effective organizations for data that can be ordered on a linear scale. The preceding chapters have exposed the most frequently used ingenious schemes for efficient searching and maintenance (insertion, deletion). Trees, however, do not seem to be helpful in problems where the data are located not in a one-dimensional, but in a multi-dimensional space. In fact, efficient searching in multi-dimensional spaces is still one of the more elusive problems in computer science, the case of two dimensions being of particular importance to many practical applications.

Upon closer inspection of the subject, trees might still be applied usefully at least in the two-dimensional case. After all, we draw trees on paper in a two-dimensional space. Let us therefore briefly review the characteristics of the two major kinds of trees so far encountered.

1. A *search tree* is governed by the invariants

$$\begin{array}{ll} p.left \neq NIL & \rightarrow p.left.x < p.x \\ p.right \neq NIL & \rightarrow p.x < p.right.x \end{array} \qquad (4.88)$$

holding for all nodes p with key x. It is apparent that only the *horizontal* position of nodes is at all constrained by the invariant, and that the vertical positions of nodes can be arbitrarily chosen such that access times in searching, (i.e. path lengths) are minimized.

2. A *heap,* also called *priority tree,* is governed by the invariants

$$\begin{array}{ll} p.left \neq NIL & \rightarrow p.y \leq p.left.y \\ p.right \neq NIL & \rightarrow p.y \leq p.right.y \end{array} \qquad (4.89)$$

holding for all nodes p with key y. Here evidently only the *vertical* positions are constrained by the invariants.

It seems straightforward to combine these two conditions in a definition of a tree organization in a two-dimensional space, with each node having two keys x and y, which can be regarded as coordinates of the node. Such a tree represents a point set in a plane, i.e. in a two-dimensional Cartesian space; it is therefore called *Cartesian tree* [4-9]. We prefer the term *priority search tree,* because it exhibits that this structure emerged from a combination of the priority tree and the search tree. It is characterized by the following invariants holding for each node p:

$$\begin{array}{ll} p.left \neq NIL & \rightarrow (p.left.x < p.x) \,\&\, (p.y \leq p.left.y) \\ p.right \neq NIL & \rightarrow (p.x < p.right.x) \,\&\, (p.y \leq p.right.y) \end{array} \qquad (4.90)$$

It should come as no big surprise, however, that the search properties of such trees are not particularly wonderful. After all, a considerable degree of freedom in positioning nodes has been taken away and is no longer available for choosing arrangements yielding short path lengths. Indeed, no logarithmic bounds on efforts involved in searching, inserting, or deleting elements can be assured. Although this had already been the case for the ordinary, unbalanced search tree, the chances for good average behaviour are slim. Even worse, maintenance operations can become rather unwieldy. Consider, for example, the tree of Fig. 4.54 (a). Insertion of a new node C whose coordinates force it to be inserted above and between A and B requires a considerable effort transforming (a) into (b).

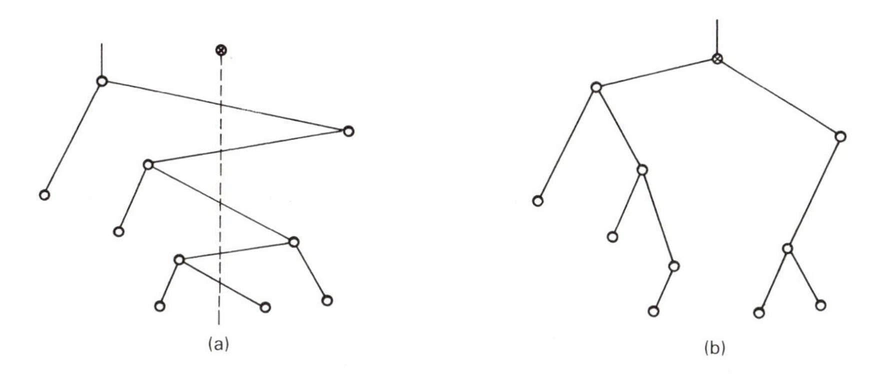

Fig. 4.54 Insertion in a priority search tree.

McCreight discovered a scheme, similar to balancing, that, at the expense of a more complicated insertion and deletion operation, guarantees logarithmic time bounds for these operations. He calls that structure a priority search tree [4-10]; in terms of our classification, however, it should be called a *balanced priority search tree.* We refrain from discussing that structure, because the scheme is very intricate and in practice hardly used. By considering a somewhat more restricted, but in practice no less relevant problem, McCreight arrived at yet another tree structure, which shall be presented here in detail. Instead of assuming that the search space be unbounded, he considered the data space to be delimited by a rectangle with two sides open. We denote the limiting values of the x-coordinate by *xmin* and *xmax.*

In the scheme of the (unbalanced) priority search tree outlined above, each node p divides the plane into two parts along the line x = p.x. All nodes of the left subtree lie to its left, all those in the right subtree to its right. For the efficiency of searching this choice may be bad. Fortunately, we may choose the dividing line differently. Let us associate with each node p an interval [p.L .. p.R), ranging over all x values including x.L up to but excluding x.R. This shall be the interval within which the x-value of the node may lie. Then we postulate that the left descendant (if any) must lie within the left half, the right descendant within the right half of this interval. Hence, the dividing line is not p.x, but (p.L+p.R)/2. For each descendant the interval is halved, thus limiting the height of the tree to $\log(x_{max}-x_{min})$. This result holds only if no two nodes have the same x-value, a condition which, however, is guaranteed by the invariant (4.90). If we deal with integer coordinates, this limit is at most equal to the wordlength of the computer used. Effectively, the search proceeds like a bisection or radix search, and therefore these trees are called *radix priority search trees* [4-10]. They feature *logarithmic bounds* on the number of operations required for searching, inserting, and deleting an element, and are governed by the following invariants for each node p:

$$\begin{array}{lll} \text{p.left} \neq \text{NIL} & \rightarrow & (\text{p.L} \leq \text{p.left.x} < \text{p.M}) \ \& \ (\text{p.y} \leq \text{p.left.y}) \\ \text{p.right} \neq \text{NIL} & \rightarrow & (\text{p.M} \leq \text{p.right.x} < \text{p.R}) \ \& \ (\text{p.y} \leq \text{p.right.y}) \end{array} \quad (4.91)$$

where

$$\begin{array}{ll} \text{p.M} & = (\text{p.L} + \text{p.R}) \text{ DIV } 2 \\ \text{p.left.L} & = \text{p.L} \\ \text{p.left.R} & = \text{p.M} \\ \text{p.right.L} & = \text{p.M} \\ \text{p.right.R} & = \text{p.R} \end{array}$$

for all node p, and root.L = x_{min}, root.R = x_{max}.

A decisive advantage of the radix scheme is that maintenance operations (preserving the invariants under insertion and deletion) are confined to a single spine of the tree, because the dividing lines have fixed values of x irrespective of the x-values of the inserted nodes.

Typical operations on priority search trees are insertion, deletion, finding an element with the least (largest) value of x (or y) larger (smaller) than a given limit, and enumerating the points lying within a given rectangle. Given below are procedures for inserting and enumerating. They are based on the following type declarations:

```
TYPE Ptr = POINTER TO Node;
  Node =  RECORD                                          (4.92)
            x: [xmin .. xmax];  y: CARDINAL;
            left, right: Ptr
          END
```

Notice that the attributes xL and xR need not be recorded in the nodes themselves. They are rather computed during each search. This, however, requires two additional parameters of the recursive procedure *insert*. Their values for the first call (with p = root) are x_{min} and x_{max} respectively. Apart from this, a search proceeds similarly to that of a regular search tree. If an empty node is encountered, the element is inserted. If the node to be inserted has a y-value smaller than the one being inspected, the new node is exchanged with the inspected node. Finally, the node inserted in the left subtree, if its x-value is less than the middle value of the interval, or the right subtree otherwise.

```
PROCEDURE insert(VAR p: Ptr; X, Y, xL, xR: CARDINAL);
 VAR xm, t: CARDINAL;
BEGIN
 IF p = NIL THEN (*not in tree, insert*)
  ALLOCATE(p, SIZE(Node));
  WITH p↑ DO
   x := X; y := Y; left := NIL; right := NIL
  END
 ELSIF p↑.x = X THEN (*found; don't insert*)
 ELSE
  IF p↑.y > Y THEN
   t := p↑.x; p↑.x := X; X := t;
```

```
      t := p↑.y; p↑.y := Y; Y := t
    END ;
    xm := (xL + xR) DIV 2;
    IF X < xm THEN insert(p↑.left, X, Y, xL, xm)
    ELSE insert(p↑.right, X, Y, xm, xR)
    END
  END
END insert
```

The task of enumerating all points x,y lying in a given rectangle, i.e. satisfying $x0 \leq x < x1$ and $y \leq y1$ is accomplished by the following procedure *enumerate*. It calls a procedure *report(x,y)* for each point found. Note that one side of the rectangle lies on the x-axis, i.e. the lower bound for y is 0. This guarantees that enumeration requires at most $O(\log N + s)$ operations, where N is the cardinality of the search space in x and s is the number of nodes enumerated.

```
PROCEDURE enumerate(p: Ptr; x0, x1, y, xL, xR: CARDINAL);
  VAR xm: CARDINAL;
BEGIN
  IF p # NIL THEN
    IF (p↑.y <= y) & (x0 <= p↑.x) & (p↑.x < x1) THEN
      report(p↑.x, p↑.y)
    END ;
    xm := (xL + xR) DIV 2;
    IF x0 < xm THEN enumerate(p↑.left, x0, x1, y, xL, xm) END ;
    IF xm < x1 THEN enumerate(p↑.right, x0, x1, y, xm, xR) END
  END
END enumerate
```

EXERCISES

4.1. Let us introduce the notion of a *recursive type*, to be declared as

```
RECTYPE T = T0
```

and denoting the set of values defined by the type T0 enlarged by the single value NONE.

The definition of the type *ped* [see(4.3)], for example, could then be simplified to

```
RECTYPE ped = RECORD name: alfa;
                 father, mother: ped
               END
```

Which is the storage pattern of the recursive structure corresponding to Fig. 4.2? Presumably, an implementation of such a feature would be based on a dynamic storage allocation scheme, and the fields named *father* and *mother* in the above example would contain pointers generated automatically but hidden from the programmer. What are the difficulties encountered in the realization of such a feature?

4.2 Define the data structure described in the last paragraph of Section 4.2 in terms of records and pointers. Is it also possible to represent this family constellation in terms of recursive types as proposed in the preceding exercise?

4.3. Assume that a first-in-first-out (fifo) queue Q with elements of type T0 is implemented as a linked list. Define a module with a suitable data structure, procedures to insert and extract an element from Q, and a function to test whether or not the queue is empty. The procedures should contain their own mechanism for an economical reuse of storage.

4.4. Assume that the records of a linked list contain a key field of type INTEGER. Write a program to sort the list in order of increasing value of the keys. Then construct a procedure to invert the list.

4.5. Circular lists (see Fig. 4.55) are usually set up with a so-called *list header.* What is the reason for introducing such a header? Write procedures for the insertion, deletion, and search of an element identified by a given key. Do this once assuming the existence of a header, once without header.

4.6. A *bidirectional list* is a list of elements that are linked in both ways. (See Fig. 4.56) Both links are originating from a header. Analogous to the preceding exercise, construct a module with procedures for searching, inserting, and deleting elements.

4.7. Does Program 4.2 work correctly if a certain pair <x,y> occurs more than once in the input?

4.8. The message *"This set is not partially ordered"* in Program 4.2 is not very helpful in many cases. Extend the program so that it outputs a sequence of elements that form a loop,

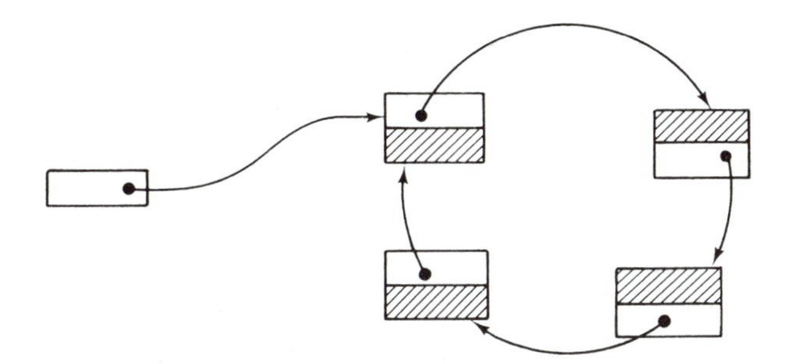

Fig. 4.55 Circular list.

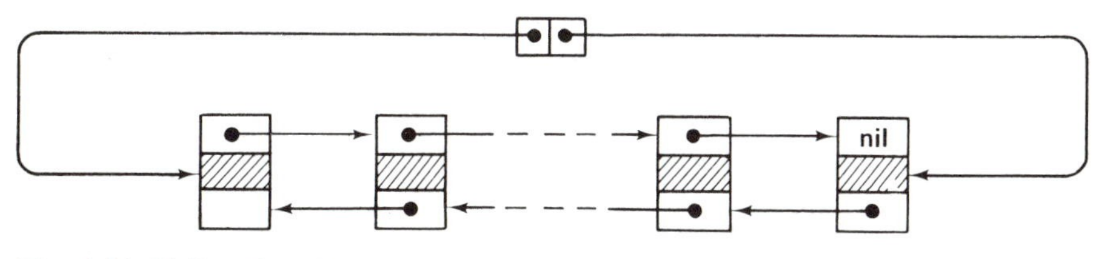

Fig. 4.56 Bidirectional list.

if there exists one.

4.9. Write a program that reads a program text, identifies all procedure definitions and calls, and tries to establish a topological ordering among the subroutines. Let $P < Q$ mean that P is called by Q.

4.10. Draw the tree constructed by Program 4.3, if the input consists of the natural numbers 1, 2, 3, ... , n.

4.11. Which are the sequences of nodes encountered when traversing the tree of Fig. 4.23 in preorder, inorder, and postorder?

4.12. Find a composition rule for the sequence of n numbers which, if applied to Program 4.4, yields a perfectly balanced tree.

4.13. Consider the following two orders for traversing binary trees:

a1. Traverse the right subtree.
a2. Visit the root.
a3. Traverse the left subtree.

b1. Visit the root.
b2. Traverse the right subtree.
b3. Traverse the left subtree.

Are there any simple relationships between the sequences of nodes encountered following these orders and those generated by the three orders defined in the text?

4.14. Define a data structure to represent n-ary trees. Then write a procedure that traverses the n-ary tree and generates a binary tree containing the same elements. Assume that the key stored in an element occupies k words and that each pointer occupies one word of storage. What is the gain in storage when using a binary tree versus an n-ary tree?

4.15. Assume that a tree is built upon the following definition of a recursive data structure (see Exercise 4.1). Formulate a procedure to find an element with a given key x and to perform an operation P on this element.

```
RECTYPE Tree =RECORD x: INTEGER;
                     left, right: Tree
                   END
```

4.16. In a certain file system a directory of all files is organized as an ordered binary tree. Each node denotes a file and specifies the file name and, among other things the date of its last access, encoded as an integer. Write a program that traverses the tree and deletes all files whose last access was before a certain date.

4.17. In a tree structure the frequency of access of each element is measured empirically by attributing to each node an access count. At certain intervals of time, the tree organization is updated by traversing the tree and generating a new tree by using Program 4.4, and inserting the keys in the order of decreasing frequency count. Write a program that performs this reorganization. Is the average path length of this tree equal to, worse, or much worse than that of an optimal tree?

4.18. The method of analyzing the tree insertion algorithm described in Sect. 4.5 can also be used to compute the expected numbers C_n of comparisons and M_n of moves (exchanges) which are performed by Quicksort (Program 2.10) sorting n elements of an array, assuming that all n! permutations of the n keys 1, 2, ... , n are equally likely. Find the analogy and determine C_n and M_n.

4.19. Draw the balanced tree with 12 nodes which has the maximum height of all 12-node balanced trees. In which sequence do the nodes have to be inserted so that procedure (4.63) generates this tree?

4.20. Find a sequence of n insertion keys so that procedure (4.63) performs each of the four rebalancing acts (LL, LR, RR, RL) at least once. What is the minimal length n for such a sequence?

4.21. Find a balanced tree with keys 1 ... n and a permutation of these keys so that, when applied to the deletion procedure (4.64), this procedure performs each of the four rebalancing routines at least once. What is the sequence with minimal length n?

4.22. What is the average path length of the Fibonacci-tree T_n?

4.23. Write a program that generates a nearly optimal tree according to the algorithm based on the selection of a centroid as root (4.78).

4.24. Assume that the keys 1, 2, 3, ... are inserted into an empty B-tree of order 2 (Program 4.7). Which keys cause page splits to occur? Which keys cause the height of the tree to increase? If the keys are deleted in the same order, which keys cause pages to be merged (and disposed) and which keys cause the height to decrease? Answer the question for (a) a deletion scheme using balancing (as in Program 4.7), and (b) a scheme whithout balancing (upon underflow, a single item only is fetched from a neighboring page).

4.25. Write a program for the search, insertion, and deletion of keys in a binary B-tree. Use the node type definition (4.84). The insertion scheme is shown in Fig. 4.51.

4.26. Find a sequence of insertion keys which, starting from the empty symmetric binary B-tree, causes procedure (4.87) to perform all four rebalancing acts (LL, LR, RR, RL) at least once. What is the shortest such sequence?

4.27. Write a procedure for the deletion of elements in a symmetric binary B-tree. Then find a tree and a short sequence of deletions causing all four rebalancing situations to occur at least once.

4.28 Formulate a data structure and procedures for the insertion and deletion of an element in a priority search tree. The procedures must maintain the invariants (4.90). Compare their performance with that of the radix priority search tree.

4.28 Design a module with the following procedures operating on radix priority search trees:
-- insert a point with coordinates x, y.
-- enumerate all points within a specified rectangle.
-- find the point with the least x-coordinate in a specified rectangle.
-- find the point with the largest y-coordinate within a specified rectangle.
-- enumerate all points lying within two (intersecting) rectangles.

REFERENCES

4-1. G.M. Adelson-Velskii and E.M. Landis. *Doklady Akademia Nauk SSSR,* 146, (1962), 263-66; English translation in *Soviet Math,* 3, 1259-63.

4-2. R. Bayer and E.M. McCreight. Organization and Maintenance of Large Ordered Indexes. *Acta Informatica, 1,* No. 3 (1972), 173-89.

4-3. -----, Binary B-trees for Virtual memory. Proc. 1971 ACM SIGFIDET Workshop, San Diego, Nov. 1971, pp. 219-35.

4-4. -----, Symmetric Binary B-trees: Data Structure and Maintenance Algorithms. *Acta Informatica, 1,* No. 4 (1972), 290-306.

4-5. T.C. Hu and A.C. Tucker. *SIAM J. Applied Math,* 21, No. 4 (1971) 514-32.

4-6. D. E. Knuth. Optimum Binary Search Trees. *Acta Informatica, 1,* No. 1 (1971), 14-25.

4-7. W.A. Walker and C.C. Gotlieb. A Top-down Algorithm for Constructing Nearly Optimal Lexicographic Trees. in *Graph Theory and Computing* (New York: Academic Press, 1972), pp. 303-23.

4-8. D. Comer. The ubiquitous B-Tree. *ACM Comp. Surveys, 11,* 2 (June 1979), 121-137.

4-9. J. Vuillemin. A unifying look at data structures. *Comm. ACM, 23,* 4 (April 1980), 229-239.

4-10. E.M. McCreight. Priority search trees. *SIAM J. of Comp.* (May 1985)

5 KEY TRANSFORMATIONS (HASHING)

5.1. INTRODUCTION

The principal question discussed in Chap. 4 at length is the following: Given a set of items characterized by a key (upon which an ordering relation is defined), how is the set to be organized so that retrieval of an item with a given key involves as little effort as possible? Clearly, in a computer store each item is ultimately accessed by specifying a storage address. Hence, the stated problem is essentially one of finding an appropriate mapping H of keys (K) into addresses (A):

$$H: K \rightarrow A$$

In Chap. 4 this mapping was implemented in the form of various list and tree search algorithms based on different underlying data organizations. Here we present yet another approach that is basically simple and very efficient in many cases. The fact that it also has some disadvantages is discussed subsequently.

The data organization used in this technique is the array structure. H is therefore a mapping transforming keys into array indices, which is the reason for the term *key transformation* that is generally used for this technique. It should be noted that we shall not need to rely on any dynamic allocation procedures; the array is one of the fundamental, static structures. The method of key transformations is often used in problem areas where tree structures are comparable competitors.

The fundamental difficulty in using a key transformation is that the set of possible key values is much larger than the set of available store addresses (array indices). Take for example names consisting of up to 16 letters as keys identifying individuals in a set of a thousand persons. Hence, there are 26^{16} possible keys which are to be mapped onto 10^3 possible indices. The function H is therefore obviously a many-to-one function. Given a key k, the first step in a retrieval (search) operation is to compute its associated index h = H(k), and the second -- evidently necessary -- step is to verify whether or not the item with the key k is indeed identified by h in the array (table) T, i.e., to check whether T[H(k)].key

= k. We are immediately confronted with two questions:

1. What kind of function H should be used?
2. How do we cope with the situation that H does not yield the location of the desired item?

The answer to the second question is that some method must be used to yield an alternative location, say index h', and, if this is still not the location of the wanted item, yet a third index h", and so on. The case in which a key other than the desired one is at the identified location is called a *collision*; the task of generating alternative indices is termed *collision handling*. In the following we shall discuss the choice of a transformation function and methods of collision handling.

5.2. CHOICE OF A HASH FUNCTION

A prerequisite of a good transformation function is that it distributes the keys as evenly as possible over the range of index values. Apart from satisfying this requirement, the distribution is not bound to any pattern, and it is actually desirable that it give the impression of being entirely random. This property has given this method the somewhat unscientific name *hashing*, i.e., chopping the argument up, or making a mess. H is called the *hash function*. Clearly, it should be efficiently computable, i.e., be composed of very few basic arithmetic operations.

Assume that a transfer function ORD(k) is avilable and denotes the ordinal number of the key k in the set of all possible keys. Assume, furthermore, that the array indices i range over the intergers 0 .. N-1, where N is the size of the array. Then an obvious choice is

$$H(k) = ORD(k) \text{ MOD } N \qquad (5.1)$$

It has the property that the key values are spread evenly over the index range, and it is therefore the basis of most key transformations. It is also very efficiently computable, if N is a power of 2. But it is exactly this case that must be avoided, if the keys are sequences of letters. The assumption that all keys are equally likely is in this case mistaken. In fact, words that differ by only a few characters then most likely map onto identical indices, thus effectively causing a most uneven distribution. It is therefore particularly recommended to let N be a prime number [5-2]. This has the conseqeunce that a full division operation is needed that cannot be replaced by a mere masking of binary digits, but this is no serious drawback on most modern computers that feature a built-in division instruction.

Often, hash funtions are used which consist of applying logical operations such as the *exclusive or* to some parts of the key represented as a sequence of binary digits. These operations may be faster than division on some computers, but they sometimes fail spectacularly to distribute the keys evenly over the range of indices. We therefore refrain from discussing such methods in further detail.

5.3. COLLISION HANDLING

If an entry in the table corresponding to a given key turns out not to be the desired item, then a collision is present, i.e., two items have keys mapping onto the same index. A second probe is necessary, one based on an index obtained in a deterministic manner from the given key. There exist several methods of generating secondary indices. An obvious one is to link all entries with identical primary index H(k) together in a linked list. This is called *direct chaining.* The elements of this list may be in the primary table or not; in the latter case, storage in which they are allocated is usually called an *overflow area.* This method has the disadvantage that secondary lists must be maintained, and that each entry must provide space for a pointer (or index) to its list of collided items.

An alternative solution for resolving collisions is to dispense with links entirely and instead simply look at other entries in the same table until the item is found or an open position is encountered, in which case one may assume that the specified key is not present in the table. This method is called *open addressing* [5-3]. Naturally, the sequence of indices of secondary probes must always be the same for a given key. The algorithm for a table lookup can then be sketched as follows:

```
h := H(k); i := 0;
REPEAT                                                    (5.2)
  IF T[h].key = k THEN item found
  ELSIF T[h].key = free THEN item is not in table
  ELSE (*collision*)
    i := i+1; h := H(k) + G(i)
  END
UNTIL found or not in table (or table full)
```

Various functions for resolving collisions have been proposed in the literature. A survey of the topic by Morris in 1968 [4-8] stimulated considerable activities in this field. The simplest method is to try for the next location -- considering the table to be circular -- until either the item with the specified key is found or an empty location is encountered. Hence, G(i) = i; the indices h_i used for probing in this case are

$$h_0 = H(k)$$
$$h_i = (h_0 + i) \text{ MOD } N, \qquad i = 1 \dots N-1 \qquad (5.3)$$

This method is called *linear probing* and has the disadvantage that entries have a tendency to cluster around the primary keys (keys that had not collided upon insertion). Ideally, of course, a function G should be chosen that again spreads the keys uniformly over the remaining set of locations. In practice, however, this tends to be too costly, and methods that offer a compromise by being simple to compute and still superior to the linear function (5.3) are preferred. One of them consists of using a quadratic function such that the sequence of indices for probing is

$$h_0 = H(k)$$

$$h_i = (h_0 + i^2) \text{ MOD } N \qquad i > 0 \qquad (5.4)$$

Note that computation of the next index need not involve the operation of squaring, if we use the recurrence relations (5.5) for $h_i = i^2$ and $d_i = 2i + 1$.

$$h_{i+1} = h_i + d_i$$
$$d_{i+1} = d_i + 2 \qquad (i > 0) \qquad (5.5)$$

with $h_0 = 0$ and $d_0 = 1$. This is called *quadratic probing*, and it essentially avoids primary clustering, although practically no additional computations are required. A very slight disadvantage is that in probing not all table entries are searched, that is, upon insertion one may not encounter a free slot although there are some left. In fact, in quadratic probing at least half the table is visited if its size N is a prime number. This assertion can be derived from the following deliberation. If the i th and the j th probes coincide upon the same table entry, we can express this by the equation

$$i^2 \text{ MOD } N = j^2 \text{ MOD } N$$
$$(i^2 - j^2) \equiv 0 \qquad (\text{modulo } N)$$

Splitting the differences up into two factors, we obtain

$$(i + j)(i - j) \equiv 0 \quad (\text{modulo } N)$$

and since $i \neq j$, we realize that either i or j have to be at least N/2 in order to yield $i+j = cN$, with c being an integer. In practice, the drawback is of no importance, since having to perform N/2 secondary probes and collision evasions is extremely rare and occurs only if the table is already almost full.

As an application of the scatter storage technique, the Cross-Reference Generator Program 4.5 is rewritten in the form of Program 5.1. The principal differences lie in the procedure *search* and in the replacement of the pointer type *WPtr* by the table of words T. The hash function H is the modulus of the table size; quadratic probing was chosen for collision handling. Note that it is essential for good performance that the table size be a prime number.

Although the method of key transformation is most effective in this case -- actually more efficient than tree organizations -- it also has a disadvantage. After having scanned the text and collected the words, we presumably wish to tabulate these words in alphabetical order. This is straightforward when using a tree organization, because its very basis is the *ordered* search tree. It is not, however, when key transformations are used. The full significance of the word *hashing* becomes apparent. Not only would the table printout have to be preceded by a sort process (which is omitted from Program 5.1), but it even turns out to be advantageous to keep track of inserted keys by linking them together explicitly in a list. Hence, the superior performance of the hashing method considering the process of retrieval only is partly offset by additional operations required to complete the full task of generating an ordered cross-reference index.

```
MODULE XRef;
```

```
FROM InOut IMPORT OpenInput, OpenOutput, CloseInput, CloseOutput,
   Read, Done, EOL, Write, WriteCard, WriteString, WriteLn;
FROM Storage IMPORT ALLOCATE;

CONST P = 997; (*prime, table size*)
   BufLeng = 10000; WordLeng = 16;
   free = 0;

TYPE WordInx = [0 .. P-1];
   ItemPtr = POINTER TO Item;

   Word = RECORD key: CARDINAL;
         first, last: ItemPtr;
      END ;

   Item = RECORD lno: CARDINAL;
         next: ItemPtr
      END ;

VAR  k0, k1, line: CARDINAL;
   ch: CHAR;
   T:  ARRAY [0 .. P-1] OF Word;  (*hash table*)
   buffer: ARRAY [0 .. BufLeng-1] OF CHAR;

PROCEDURE PrintWord(k: CARDINAL);
  VAR lim: CARDINAL;
BEGIN lim := k + WordLeng;
  WHILE buffer[k] > 0C DO Write(buffer[k]); k := k+1 END ;
  WHILE k < lim DO Write(" "); k := k+1 END
END PrintWord;

PROCEDURE PrintTable;
  VAR i, k, m: CARDINAL; item: ItemPtr;
BEGIN
  FOR k := 0 TO P-1 DO
    IF T[k].key # free THEN
      PrintWord(T[k].key); item := T[k].first; m := 0;
      REPEAT
        IF m = 8 THEN
          WriteLn; m := 0;
          FOR i := 1 TO WordLeng DO Write(" ") END
        END ;
        m := m+1; WriteCard(item↑.lno, 6); item := item↑.next
      UNTIL item = NIL;
      WriteLn;
    END
  END
END PrintTable;
```

```
PROCEDURE Diff(i, j: CARDINAL): INTEGER;
BEGIN
  LOOP
    IF buffer[i] # buffer[j] THEN
      RETURN INTEGER(ORD(buffer[i])) - INTEGER(ORD(buffer[j]))
    ELSIF buffer[i] = 0C THEN RETURN 0
    END ;
    i := i+1; j := j+1
  END
END Diff;

PROCEDURE search;
  VAR i, h, d: CARDINAL; found: BOOLEAN;
    ch: CHAR; x: ItemPtr;
  (*global variables: T, buffer, k0, k1*)
BEGIN (*compute hash index h for word starting at buffer[k0]*)
  i := k0; h := 0; ch := buffer[i];
  WHILE ch > 0C DO
    h := (256*h + ORD(ch)) MOD P; i := i+1; ch := buffer[i]
  END ;
  ALLOCATE(x, SIZE(Item)); x↑.lno := line; x↑.next := NIL;
  d := 1; found := FALSE;
  REPEAT
   IF Diff(T[h].key, k0) = 0 THEN (*match*)
    found := TRUE; T[h].last↑.next := x; T[h].last := x
   ELSIF T[h].key = free THEN (*new entry*)
    WITH T[h] DO
      key := k0; first := x; last := x
    END ;
    found := TRUE; k0 := k1
   ELSE (*collision*) h := h+d; d := d+2;
    IF h >= P THEN h := h-P END ;
    IF d = P THEN WriteString(" Table overflow"); HALT END
   END
  UNTIL found
END search;

PROCEDURE GetWord;
BEGIN k1 := k0;
  REPEAT Write(ch); buffer[k1] := ch; k1 := k1 + 1; Read(ch)
  UNTIL (ch < "0") OR (ch > "9") & (CAP(ch) < "A")
    OR (CAP(ch) > "Z");
  buffer[k1] := 0C; k1 := k1 + 1; (*terminator*)
  search
END GetWord;
```

```
BEGIN k0 := 1; line := 0;
  FOR k1 := 0 TO P-1 DO T[k1].key := free END ;
  OpenInput("TEXT"); OpenOutput("XREF");
  WriteCard(0, 6); Write(" "); Read(ch);
  WHILE Done DO
    CASE ch OF
      0C .. 35C:   Read(ch) |
      36C .. 37C:  WriteLn; Read(ch); line := line + 1;
                   WriteCard(line, 6); Write(" ") |
      " " .. "@":  Write(ch); Read(ch) |
      "A" .. "Z":  GetWord |
      "[" .. "‘":  Write(ch); Read(ch) |
      "a" .. "z":  GetWord |
      "{" .. "~":  Write(ch); Read(ch)
    END
  END ;
  WriteLn; WriteLn; CloseInput;
  PrintTable; CloseOutput
END XRef.
```

Program 5.1 Cross Reference Generator Using Hash Table

5.4. ANALYSIS OF KEY TRANSFORMATION

Insertion and retrieval by key transformation has evidently a miserable worst-case performance. After all, it is entirely possible that a search argument may be such that the probes hit exactly all occupied locations, missing consistently the desired (or free) ones. Actually, considerable confidence in the correctness of the laws of probability theory is needed by anyone using the hash technique. What we wish to be assured of is that on the *average* the number of probes is small. The following probabilistic argument reveals that it is even very small.

Let us once again assume that all possible keys are equally likely and that the hash function H distributes them uniformly over the range of table indices. Assume, then, that a key has to be inserted in a table of size n which already contains k items. The probability of hitting a free location the first time is then (n-k)/n. This is also the probability p_1 that a single comparison only is needed. The probability that excatly one second probe is needed is equal to the probability of a collision in the first try times the probability of hitting a free location the next time. In general, we obtain the probability p_i of an insertion requiring exactly i probes as

$$
\begin{aligned}
p_1 &= (n-k)/n \\
p_2 &= (k/n) * (n-k)/(n-1) \\
p_3 &= (k/n) * (k-1)/(n-1) * (n-k)/(n-2) \\
&\dots \\
p_i &= (k/n) * (k-1)/(n-1) * (k-2)/(n-2) * \dots * (n-k)/(n-i+1)
\end{aligned}
\tag{5.6}
$$

The expected number E of probes required upon insertion of the k+1st key is therefore

$$
\begin{aligned}
E_{k+1} &= Si : 1 \le i \le k+1 : i*p_i \\
&= 1 * (n-k)/n \; + \; 2 * (k/n) * (n-k)/(n-1) \; + \; \dots \; + \\
&\quad (k+1) * (k/n) * (k-1)/(n-1) * (k-2)/(n-2) * \dots * 1/(n-k+1) \\
&= (n+1)/(n-k+1)
\end{aligned}
\tag{5.7}
$$

Since the number of probes required to insert an item is identical with the number of probes needed to retrieve it, the result (4.94) can be used to compute the average number E of probes needed to access a random key in a table. Let the table size again be denoted by n, and let m be the number of keys actually in the table. Then

$$
\begin{aligned}
E &= (Sk : 1 \le k \le m : E_k)/m \\
&= (n+1) * (Sk : 1 \le k \le m : 1/(n-k+2))/m \\
&= (n+1) * (H_{n+1} - H_{n-m+1})/m
\end{aligned}
\tag{5.8}
$$

where H is the harmonic function. H can be approximated as $H_n = \ln(n) + g$, where g is Euler's constant. If, moreover, we substitute $a = m/(n+1)$, we obtain

$$
\begin{aligned}
E &= (\ln(n+1) - \ln(n-m+1))/a = \ln((n+1)/(n-m+1))/a \\
&= -\ln(1-a)/a
\end{aligned}
\tag{5.9}
$$

a is approximately the quotient of occupied and available locations, called the *load factor*; a = 0 implies an empty table, a = n/(n+1) a full table. The expected number E of probes to retrieve or insert a randomly chosen key is listed in Table 5.1 as a function of the load factor. The numerical results are indeed surprising, and they explain the exceptionally good performance of the key transformation method. Even if a table is 90% full, on the average only 2.56 probes are necessary to either locate the key or to find an empty location. Note in particular that this figure does not depend on the absolute number of keys present, but only on the load factor.

a	E
0.1	1.05
0.25	1.15
0.5	1.39
0.75	1.85
0.9	2.56
0.95	3.15
0.99	4.66

Table 4.6 Expected number of probes as a function of the load factor.

The above analysis was based on the use of a collision handling method that spreads the keys uniformly over the remaining locations. Methods used in practice yield slightly worse performance. Detailed analysis for linear probing yields an expected number of probes as

$$E \quad = (1 - a/2)/(1 - a) \tag{5.10}$$

Some numerical values for E(a) are listed in Table 4.7 [5-4]. The results obtained even for the poorest method of collision handling are so good that there is a temptation to regard key transformation (hashing) as the panacea for everything. This is particularly so because its performance is superior even to the most sophisticated tree organization discussed, at least on the basis of comparison steps needed for retrieval and insertion. It is therefore important to point out explicitly some of the drawbacks of hashing, even if they are obvious upon unbiased consideration.

a	E
0.1	1.06
0.25	1.17
0.5	1.50
0.75	2.50
0.9	5.50
0.95	10.50

Table 4.7 Expected number of probes for linear probing.

Certainly the major disadvantage over techniques using dynamic allocation is that the size of the table is fixed and cannot be adjusted to actual demand. A fairly good *a priori* estimate of the number of data items to be classified is therefore mandatory if either poor storage utilization or poor performance (or even table overflow) is to be avoided. Even if the number of items is exactly known -- an extremely rare case -- the desire for good performance dictates to dimension the table slightly (say 10%) too large.

The second major deficiency of scatter storage techniques becomes evident if keys are not only to be inserted and retrieved, but if they are also to be deleted. Deletion of entries in a hash table is extremely cumbersome unless direct chaining in a separate overflow area is used. It is thus fair to say that tree organizations are still attractive, and actually to be preferred, if the volume of data is largely unknown, is strongly variable, and at times even decreases.

EXERCISES

5.1. If the amount of information associated with each key is relatively large (compared to the key itself), this information should not be stored in the hash table. Explain why and propose a scheme for representing such a set of data.

5.2. Consider the proposal to solve the clustering problem by constructing overflow trees instead of overflow lists, i.e., of organizing those keys that collided as tree structures. Hence, each entry of the scatter (hash) table can be considered as the root of a (possibly empty) tree. Compare the expected performance of this tree hashing method with that of open addressing.

5.3. Devise a scheme that performs insertions and deletions in a hash table using quadratic increments for collision resolution. Compare this scheme experimentally with the straight binary tree organization by applying random sequences of keys for insertion and deletion.

5.4. The primary drawback of the hash table technique is that the size of the table has to be fixed at a time when the actual number of entries is not known. Assume that your computer system incorporates a dynamic storage allocation mechanism that allows to obtain storage at any time. Hence, when the hash table H is full (or nearly full), a larger table H' is generated, and all keys in H are transferred to H', whereafter the store for H can be returned to the storage administration. This is called *rehashing.* Write a program that performs a rehash of a table H of size n.

5.5. Very often keys are not integers but sequences of letters. These words may greatly vary in length, and therefore they cannot conveniently and economically be stored in key fields of fixed size. Write a program that operates with a hash table and variable length keys.

REFERENCES

5-1. W.D. Maurer. An Improved Hash Code for Scatter Storage. *Comm. ACM, 11,* No. 1 (1968), 35-38.

5-2. R. Morris. Scatter Storage Techniques. *Comm. ACM, 11,* No. 1 (1968), 38-43.

5-3. W.W. Peterson. Addressing for Random-access Storage. *IBM J. Res. & Dev., 1,* (1957), 130-46.

5-4. G. Schay and W. Spruth. Analysis of a File Addressing Method. *Comm. ACM, 5,* No. 8 (1962), 459-62.

A The ASCII Character Set

<table>
<tr><th></th><th>0</th><th>10</th><th>20</th><th>30</th><th>40</th><th>50</th><th>60</th><th>70</th></tr>
<tr><td>0</td><td>nul</td><td>dle</td><td></td><td>0</td><td>@</td><td>P</td><td>‘</td><td>p</td></tr>
<tr><td>1</td><td>soh</td><td>dc1</td><td>!</td><td>1</td><td>A</td><td>Q</td><td>a</td><td>q</td></tr>
<tr><td>2</td><td>stx</td><td>dc2</td><td>"</td><td>2</td><td>B</td><td>R</td><td>b</td><td>r</td></tr>
<tr><td>3</td><td>etx</td><td>dc3</td><td>#</td><td>3</td><td>C</td><td>S</td><td>c</td><td>s</td></tr>
<tr><td>4</td><td>eot</td><td>dc4</td><td>$</td><td>4</td><td>D</td><td>T</td><td>d</td><td>t</td></tr>
<tr><td>5</td><td>enq</td><td>nak</td><td>%</td><td>5</td><td>E</td><td>U</td><td>e</td><td>u</td></tr>
<tr><td>6</td><td>ack</td><td>syn</td><td>&</td><td>6</td><td>F</td><td>V</td><td>f</td><td>v</td></tr>
<tr><td>7</td><td>bel</td><td>etb</td><td>'</td><td>7</td><td>G</td><td>W</td><td>g</td><td>w</td></tr>
<tr><td>8</td><td>bs</td><td>can</td><td>(</td><td>8</td><td>H</td><td>X</td><td>h</td><td>x</td></tr>
<tr><td>9</td><td>ht</td><td>em</td><td>)</td><td>9</td><td>I</td><td>Y</td><td>i</td><td>y</td></tr>
<tr><td>A</td><td>lf</td><td>sub</td><td>*</td><td>:</td><td>J</td><td>Z</td><td>j</td><td>z</td></tr>
<tr><td>B</td><td>vt</td><td>esc</td><td>+</td><td>;</td><td>K</td><td>[</td><td>k</td><td>{</td></tr>
<tr><td>C</td><td>ff</td><td>fs</td><td>,</td><td><</td><td>L</td><td>\</td><td>l</td><td>|</td></tr>
<tr><td>D</td><td>cr</td><td>gs</td><td>-</td><td>=</td><td>M</td><td>]</td><td>m</td><td>}</td></tr>
<tr><td>E</td><td>so</td><td>rs</td><td>.</td><td>></td><td>N</td><td>↑</td><td>n</td><td>~</td></tr>
<tr><td>F</td><td>si</td><td>us</td><td>/</td><td>?</td><td>O</td><td>←</td><td>o</td><td>del</td></tr>
</table>

Layout characters

bs	backspace
ht	horizontal tabulator
lf	line feed
vt	vertical tabulator
ff	form feed
cr	carriage return

Separator characters

fs	file separator
gs	group separator
rs	record separator
us	unit separator

B Syntax of Modula-2

```
1   ident = letter {letter | digit}.
2   number = integer | real.
3   integer = digit {digit} | octalDigit {octalDigit} ("B"|"C")|
4     digit {hexDigit} "H".
5   real = digit {digit} "." {digit} [ScaleFactor].
6   ScaleFactor = "E" ["+"|"-"] digit {digit}.
7   hexDigit = digit |"A"|"B"|"C"|"D"|"E"|"F".
8   digit = octalDigit | "8"|"9".
9   octalDigit = "0"|"1"|"2"|"3"|"4"|"5"|"6"|"7".
10  string = "'" {character} "'" | '"' {character} '"' .
11  qualident = ident {"." ident}.
12  ConstantDeclaration = ident "=" ConstExpression.
13  ConstExpression = expression.
14  TypeDeclaration = ident "=" type.
15  type = SimpleType | ArrayType | RecordType | SetType |
16    PointerType | ProcedureType.
17  SimpleType = qualident | enumeration | SubrangeType.
18  enumeration = "(" IdentList ")".
19  IdentList = ident {"," ident}.
20  SubrangeType = [qualident] "[" ConstExpression ".." ConstExpression "]".
21  ArrayType = ARRAY SimpleType {"," SimpleType} OF type.
22  RecordType = RECORD FieldListSequence END.
23  FieldListSequence = FieldList {";" FieldList}.
24  FieldList = [IdentList ":" type |
25    CASE [ident] ":" qualident OF variant {"|" variant}
26    [ELSE FieldListSequence] END].
27  variant = [CaseLabelList ":" FieldListSequence].
28  CaseLabelList = CaseLabels {"," CaseLabels}.
29  CaseLabels = ConstExpression [".." ConstExpression].
30  SetType = SET OF SimpleType.
31  PointerType = POINTER TO type.
32  ProcedureType = PROCEDURE [FormalTypeList].
33  FormalTypeList = "(" [[VAR] FormalType
34    {"," [VAR] FormalType}] ")" [":" qualident].
35  VariableDeclaration = IdentList ":" type.
36  designator = qualident {"." ident | "[" ExpList "]" | "↑"}.
37  ExpList = expression {"," expression}.
38  expression = SimpleExpression [relation SimpleExpression].
```

```
39  relation = "=" | "#" | "<" | "<=" | ">" | ">=" | IN .
40  SimpleExpression = ["+"|"-"] term {AddOperator term}.
41  AddOperator = "+" | "-" | OR .
42  term = factor {MulOperator factor}.
43  MulOperator = "*" | "/" | DIV | REM | MOD | AND .
44  factor = number | string | set | designator [ActualParameters] |
45     "(" expression ")" | NOT factor.
46  set = [qualident] "{" [element {"," element}] "}".
47  element = ConstExpression [".." ConstExpression].
48  ActualParameters = "(" [ExpList] ")" .
49  statement = [assignment | ProcedureCall |
50     IfStatement | CaseStatement | WhileStatement |
51     RepeatStatement | LoopStatement | ForStatement |
52     WithStatement | EXIT | RETURN [expression] ].
53  assignment = designator ":=" expression.
54  ProcedureCall = designator [ActualParameters].
55  StatementSequence = statement {";" statement}.
56  IfStatement = IF expression THEN StatementSequence
57     {ELSIF expression THEN StatementSequence}
58     [ELSE StatementSequence] END.
59  CaseStatement = CASE expression OF case {"|" case}
60     [ELSE StatementSequence] END.
61  case = [CaseLabelList ":" StatementSequence].
62  WhileStatement = WHILE expression DO StatementSequence END.
63  RepeatStatement = REPEAT StatementSequence UNTIL expression.
64  ForStatement = FOR ident ":=" expression TO expression
65     [BY ConstExpression] DO StatementSequence END.
66  LoopStatement = LOOP StatementSequence END.
67  WithStatement = WITH designator DO StatementSequence END .
68  ProcedureDeclaration = ProcedureHeading ";" (block ident | FORWARD).
69  ProcedureHeading = PROCEDURE ident [FormalParameters].
70  block = {declaration} [BEGIN StatementSequence] END.
71  declaration = CONST {ConstantDeclaration ";"} |
72     TYPE {TypeDeclaration ";"} |
73     VAR {VariableDeclaration ";"} |
74     ProcedureDeclaration ";" | ModuleDeclaration ";".
75  FormalParameters =
76     "(" [FPSection {";" FPSection}] ")" [":" qualident].
77  FPSection = [VAR] IdentList ":" FormalType.
78  FormalType = [ARRAY OF] qualident.
79  ModuleDeclaration =
80     MODULE ident [priority] ";" {import} [export] block ident.
81  priority = "[" ConstExpression "]".
82  export = EXPORT [QUALIFIED] IdentList ";".
83  import = [FROM ident] IMPORT IdentList ";".
84  DefinitionModule = DEFINITION MODULE ident ";"
```

```
85        {import} {definition} END ident ".".
86    definition = CONST {ConstantDeclaration ";"} |
87        TYPE {ident ["=" type] ";"} |
88        VAR {VariableDeclaration ";"} |
89        ProcedureHeading ";" .
90    ProgramModule = MODULE ident [priority] ";" {import} block ident "." .
91    CompilationUnit = DefinitionModule | [IMPLEMENTATION] ProgramModule.
```

ActualParameters	54	-48	44							
AddOperator	-41	40								
ArrayType	-21	15								
assignment	-53	49								
block	90	80	-70	68						
case	-61	59	59							
CaseLabelList	61	-28	27							
CaseLabels	-29	28	28							
CaseStatement	-59	50								
character	10	10								
CompilationUnit	-91									
ConstantDeclaration	86	71	-12							
ConstExpression	81	65	47	47	29	29	20	20	-13	12
declaration	-71	70								
definition	-86	85								
DefinitionModule	91	-84								
designator	67	54	53	44	-36					
digit	-8	7	6	6	5	5	5	4	3	3	1
element	-47	46	46							
enumeration	-18	17								
ExpList	48	-37	36							
export	-82	80								
expression	64	64	63	62	59	57	56	53	52	
		45	-38	37	37	13				
factor	45	-44	42	42						
FieldList	-24	23	23							
FieldListSequence	27	26	-23	22						
FormalParameters	-75	69								
FormalType	-78	77	34	33						
FormalTypeList	-33	32								
ForStatement	-64	51								
FPSection	-77	76	76							
hexDigit	-7	4								
ident	90	90	87	85	84	83	80	80	69	68
	64	36	25	19	19	14	12	11	11	-1
IdentList	83	82	77	35	24	-19	18			
IfStatement	-56	50								
import	90	85	-83	80						

integer	-3	2							
letter	1	1							
LoopStatement	-66	51							
ModuleDeclaration	-79	74							
MulOperator	-43	42							
number	44	-2							
octalDigit	-9	8	3	3					
PointerType	-31	16							
priority	90	-81	80						
ProcedureCall	-54	49							
ProcedureDeclaration	74	-68							
ProcedureHeading	89	-69	68						
ProcedureType	-32	16							
ProgramModule	91	-90							
qualident	78	76	46	36	34	25	20	17	-11
real	-5	2							
RecordType	-22	15							
relation	-39	38							
RepeatStatement	-63	51							
ScaleFactor	-6	5							
set	-46	44							
SetType	-30	15							
SimpleExpression	-40	38	38						
SimpleType	30	21	21	-17	15				
statement	55	55	-49						
StatementSequence	70	67	66	65	63	62	61	60	58
	57	56	-55						
string	44	-10							
SubrangeType	-20	17							
term	-42	40	40						
type	87	35	31	24	21	-15	14		
TypeDeclaration	72	-14							
VariableDeclaration	88	73	-35						
variant	-27	25	25						
WhileStatement	-62	50							
WithStatement	-67	52							

INDEX

Index of Programs: